T0205626

Environmental Footprints and Eco-design of Products and Processes

Series Editor

Subramanian Senthilkannan Muthu, Head of Sustainability - SgT Group and API, Hong Kong, Kowloon, Hong Kong

Indexed by Scopus

This series aims to broadly cover all the aspects related to environmental assessment of products, development of environmental and ecological indicators and eco-design of various products and processes. Below are the areas fall under the aims and scope of this series, but not limited to: Environmental Life Cycle Assessment; Social Life Cycle Assessment; Organizational and Product Carbon Footprints; Ecological, Energy and Water Footprints; Life cycle costing; Environmental and sustainable indicators; Environmental impact assessment methods and tools; Eco-design (sustainable design) aspects and tools; Biodegradation studies; Recycling; Solid waste management; Environmental and social audits; Green Purchasing and tools; Product environmental footprints; Environmental management standards and regulations; Eco-labels; Green Claims and green washing; Assessment of sustainability aspects.

More information about this series at http://www.springer.com/series/13340

Subramanian Senthilkannan Muthu
Editor

Life Cycle Sustainability Assessment (LCSA)

Editor
Subramanian Senthilkannan Muthu
Head of Sustainability
SgT Group and API
Kowloon, Hong Kong

ISSN 2345-7651 ISSN 2345-766X (electronic)
Environmental Footprints and Eco-design of Products and Processes
ISBN 978-981-16-4564-8 ISBN 978-981-16-4562-4 (eBook)
https://doi.org/10.1007/978-981-16-4562-4

This Springer imprint is published by the registered company Springer Nature Singapore Pte Ltd.
The registered company address is: 152 Beach Road, #21-01/04 Gateway East, Singapore 189721, Singapore

Contents

Evolution of Life Cycle Sustainability Assessment 1
Shilpi Shrivastava and Seema Unnikrishnan

A Sustainability Assessment Framework for the Australian Food Industry: Integrating Life Cycle Sustainability Assessment and Circular Economy 15
Murilo Pagotto, Anthony Halog, Diogo Fleury Azevedo Costa, and Tianchu Lu

Life Cycle Sustainability Assessment: Methodology and Framework 43
Shilpi Shrivastava and Seema Unnikrishnan

Application of Life Cycle Sustainability Assessment to Evaluate the Future Energy Crops for Sustainable Energy and Bioproducts 57
R. Anitha, R. Subashini, and P. Senthil Kumar

Sustainable Development: ICT, New Directions, and Strategies 81
Florin Dragan and Larisa Ivascu

Implementing Life Cycle Sustainability Assessment in Building and Energy Retrofit Design—An Investigation into Challenges and Opportunities 103
Hashem Amini Toosi, Monica Lavagna, Fabrizio Leonforte, Claudio Del Pero, and Niccolò Aste

Evaluating the Sustainability of Feedlot Production in Australia Using a Life Cycle Sustainability Assessment Framework 137
Murilo Pagotto, Anthony Halog, Diogo Fleury Azevedo Costa, and Tianchu Lu

Life Cycle Sustainability Assessment Study of Conventional and Prefabricated Construction Methods: MADM Analysis 179
Ali Tighnavard Balasbaneh, David Yeoh, and Mohd Irwan Juki

About the Editor

Dr. Subramanian Senthilkannan Muthu currently works for SgT Group as Head of Sustainability, and is based out of Hong Kong. He earned his Ph.D. from The Hong Kong Polytechnic University, and is a renowned expert in the areas of Environmental Sustainability in Textiles and Clothing Supply Chain, Product Life Cycle Assessment (LCA), Ecological Footprint and Product Carbon Footprint Assessment (PCF) in various industrial sectors. He has five years of industrial experience in textile manufacturing, research and development and textile testing and over a decades of experience in life cycle assessment (LCA), carbon and ecological footprints assessment of various consumer products. He has published more than 100 research publications, written numerous book chapters and authored/edited over 100 books in the areas of Carbon Footprint, Recycling, Environmental Assessment and Environmental Sustainability.

Evolution of Life Cycle Sustainability Assessment

Shilpi Shrivastava and Seema Unnikrishnan

Abstract The main objective of this chapter is to provide a brief introduction to how life cycle sustainability assessment (LCSA) evolved, explaining each dimension of sustainability assessment. The key to achieving this sustainable development goal is the protection of the environment. Environmental life cycle assessment (E-LCA) has developed fast over the last three decades. The chapter starts by explaining how E-LCA has developed from simply energy analysis to a comprehensive environmental burden analysis, and also how it has broadened the scope of LCA from studying only environmental impacts to covering all three sustainability aspects (environmental, economic, and social). Further, the analysis of the research paper focusing on sustainability assessment has been done using the keyword analysis and database search method followed by explaining each sustainability dimension which will give a brief insight of overall sustainability assessment.

Keywords LCSA · Environmental · Economic · Social · Sustainability · Sustainable development · LCA

1 Introduction

Over the last three decades, environmental life cycle assessment (E-LCA) has progressed from basic energy analysis to a detailed environmental burden analysis that includes all the dimensions of sustainability. In the year 1980s and 1990s, life cycle impact assessment, life cycle costing models, and the social-LCA gained importance in the first decade of the twenty-first century [21]. The present decade is a decade of life cycle sustainability assessment (Fig. 1). By integrating the three dimensions of sustainability, namely environment, social, and cost analysis, we can

S. Shrivastava (✉) · S. Unnikrishnan
Center for Environmental Studies, National Institute of Industrial Engineering, Mumbai, India
e-mail: shilpi.shrivastava.2016@nitie.ac.in

S. Unnikrishnan
e-mail: seemaunnikrishnan@nitie.ac.in

S. S. Muthu (ed.), *Life Cycle Sustainability Assessment (LCSA)*,
Environmental Footprints and Eco-design of Products and Processes,
https://doi.org/10.1007/978-981-16-4562-4_1

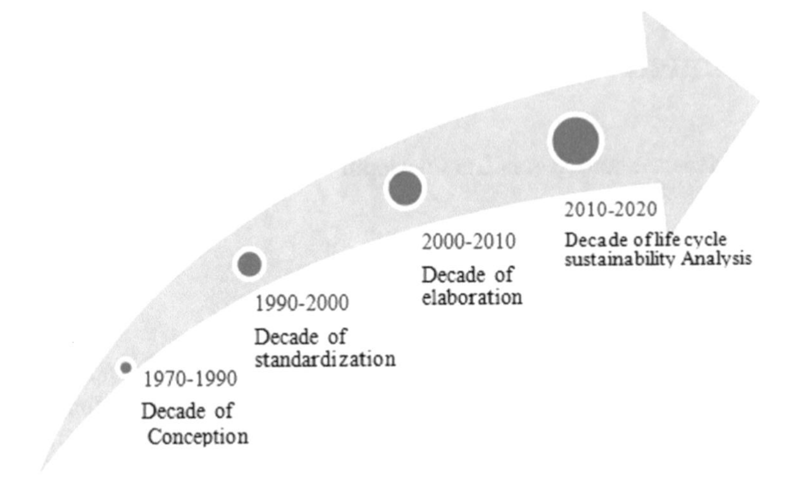

Fig. 1 Evolution of LCSA. *Source* [21]

obtain a more precise and long-term approach for life cycle analysis. However, there is still a lack of convergence of all the pillars into a shared methodological system. One explanation for this is the existence of the indicators used in the tools, which range from quantitative to qualitative, making integration challenging [73].

1.1 Decade of Conception

The decades of 1970–1990 were known as the decades of conception. During this period, there were diverse approaches for performing LCA but the international standard was missing. LCAs were implemented in various industries but by using various methods. The common theoretical framework was missing (Guinée et al. 1993). The results also varied greatly for the same study objects.

1.2 Decade of Standardization

There was a noteworthy development during this decade. Various scientific activities were initiated worldwide regarding LCA, and various LCA guidelines and LCA handbooks were published [12, 20, 25, 37]. The SETAC and ISO's standardization activities showed convergence and harmonization of methods.

1.3 Decade of Elaboration

During this decade, the attention was more on elaborating LCA, which resulted in textbooks on LCA (e.g. [5, 10, 13, 20, 36]). More research on LCA again led to the divergence of methods. New approaches regarding system boundaries and allocation methods were developed [22]. Also, specific approaches to the cost of the life cycle (LCC) and social life cycle assessment (SLCA) have been evolved. The three dimensions (or triple bottom line, TBL) model of sustainability underpins this extension of the LCA to the LCC and SLCA context, which distinguishes environmental, economic, and social impacts of product systems over their life cycle.

1.4 Decade of Life Cycle Sustainability Assessment

The decade of 2010–2020 has been designated as the life cycle sustainability assessment. During this decade, the definition of LCA was broadened from covering only environmental LCA to a more detailed life cycle sustainability assessment (LCSA), which encompassed all three dimensions of sustainability (environment, economic, and social).

2 Analysis of Paper Focusing on LCSA

Some of the key findings of selected articles focusing on LCSA are presented in Table 1.

The key flaw discovered when doing a literature review on "Life Cycle Sustainability Evaluation" is that there is no connection between the three pillars of sustainability. Selected articles were analyzed to see the present LCSA framework models developed by various authors, but it is limited to certain disciplines. Some of the other findings are elaborated in Table 2.

3 Life Cycle Attributes (Three Dimensions of Sustainability)

3.1 Environmental Life Cycle Assessment (E-LCA)

The LCA tool emerged in the 1960s and was standardized in the 1990s. Until then it was used with different names and techniques [5]. LCA is a tool used to assess the environmental implications of a product from raw material production through processing, delivery, usage, repair and maintenance, and disposal or recycling during

Table 1 Key findings of selected articles focusing on LCSA

Authors (Year)	Key findings
Luu and Halog [39]	In an effort to integrate ideas from the sustainable development model, life cycle sustainability assessment (LCSA) expands the environmental boundaries of conventional life cycle assessment
Zamagni et al. [72]	The life cycle approach is a good way to incorporate sustainability into product and service creativity, design, and evaluation
Sala et al. [54]	Improved methodologies for integrated evaluation and growing acceptance of life cycle thinking from product creation to strategic policy support are needed for progress toward sustainability
Finkbeiner et al. [18]	The paradigm shift from environmental protection to sustainability as well as current developments in evaluation methods and tools for environmental and sustainability results are the key drivers for scientific developments in life cycle sustainability assessment (LCSA)
Benedetto and Klemeš [6]	Adding further dimensions to LCA would aid in determining whether goods, practices, and services are moving in the direction of sustainability and in making strategic decisions
Zohu et al. (2006)	Working to develop tools that can accurately quantify sustainability is necessary for identifying non-sustainable practices, educating designers about product quality, and tracking social impacts

its life cycle (toward a life cycle sustainable assessment, UNEP report 2012). It is the "compilation and evaluation of inputs and outputs and the potential impacts of a product system throughout its life cycle" [27]. The ISO 14,040 series provides a detailed framework for performing the LCA. According to the ISO 14,040 series, LCA consists of four phases: goal and scope definition, inventory analysis, impact assessment, and interpretation. In the first phase of the study, system boundary, functional unit, targeted audience, and assumptions (if any) must be reported. The second step is the collection of inventory data, which includes the inputs (raw material requirement, power consumption, product, by-product, etc.) and outputs (air emissions, water emissions, waste generated, solid waste, etc.). The third step is the impact assessment phase where the environmental impact is calculated using LCA software and the last phase is the interpretation of the results.

The endpoints of a life cycle impact assessment are the impact on human health, environmental degradation, and resource depletion (Fig. 2).

By collecting an inventory of applicable energy and material inputs and environmental releases, assessing the possible impacts associated with defined inputs and releases, and analyzing the findings, LCA offers a broad view of environmental issues. LCA can help with finding ways to enhance product environmental efficiency, educating decision-makers in business, government, and non-government organizations, selecting appropriate environmental performance metrics, and marketing [33]. Also, it is wider spread than the other life cycle attributes.

In the twentieth century, the industries were very slowly recognizing the acceptance of the LCA method, but the LCA technique was gradually getting accepted.

Table 2 Analysis of selected research papers

S. no.	Authors (Year)	Sustainability pillars			Scope		Integrated framework/Model presented	
		Environmental	Economic	Social	Product-specific	Supply chain	Yes	No
1	Abu-Rayash and Dincer [1]	*	*	*	*		*	
2	Akber et al. [2]	*	*	*	*		*	
3	Akhtar et al. [3]	*	*	*	*		*	
4	Balieu et al. [4]	*				*		*
5	Chen and Holden [7]	*	*	*		*	*	
6	Contreras-Lisperguer et al. [9]	*	*	*	*			*
7	De Luca et al. [11]	*	*	*	*			*
8	Ekener et al. [14]	*	*	*	*			*
9	Elhuni et al. [15]	*	*	*		*		
10	Halog and Manik [23]	*	*	*			*	
11	Hoque et al. [26]	*	*	*	*		*	
12	Janjua et al. [30]	*	*	*	*		*	
13	Kabayo et al. [32]	*	*	*	*		*	
14	Kloepffer [34, 35]	*	*	*	*			*
15	Li et al. [38]	*	*	*		*	*	
16	Mahbub et al. [40]	*	*	*	*		*	
17	Sharma and Strezov [59]	*	*		*			*

(continued)

Table 2 (continued)

S. no.	Authors (Year)	Sustainability pillars			Scope		Integrated framework/Model presented	
		Environmental	Economic	Social	Product-specific	Supply chain	Yes	No
18	Marwa Hannouf and Getachew Assefa [24]	*	*	*		*	*	
19	Mehmeti et al. [41]	*	*		*		*	
20	Menikpura et al. [42]	*	*	*	*		*	
21	Nguyen et al. [44]	*	*	*	*		*	
22	Oliveira et al. [45]	*			*			*
23	Onat et al. [46]	*	*	*	*		*	
24	Opher et al. [47]	*	*	*		*	*	
25	Osorio-Tejada et al. [48]	*	*	*		*		*
26	Pereira and Ortega [49]	*			*			*
27	Roinioti and Koroneos [52]	*	*	*		*		*
28	Rivas Bolivar [51]	*	*	*		*		*
29	Ruiz et al. [53]	*	*	*	*		*	
30	Sala et al. [55]	*	*	*			*	
31	Santoyo-Castelazo and Azapagic [56]	*	*	*		*		*

(continued)

Table 2 (continued)

S. no.	Authors (Year)	Sustainability pillars			Scope		Integrated framework/Model presented	
		Environmental	Economic	Social	Product-specific	Supply chain	Yes	No
32	Settembre-Blundo et al. [58]	*	*	*	*			*
33	Stamford and Azapagic [60]	*	*	*	*			*
34	Ugwuoke and Oduoza [63]	*			*		*	
35	Valente et al. [65]	*	*	*	*			*
36	Vinyes et al. [66]	*	*	*		*	*	
37	Foolmaun and Ramjeawon [19]	*	*	*	*			*
38	Tye [62]	*			*			*
39	Wang et al. [67]	*	*	*	*		*	
40	Yıldız-Geyhan et al. [71]	*		*		*	*	
41	Zheng et al. [74]	*	*	*	*			*
42	Zhou et al. [75]	*	*		*		*	

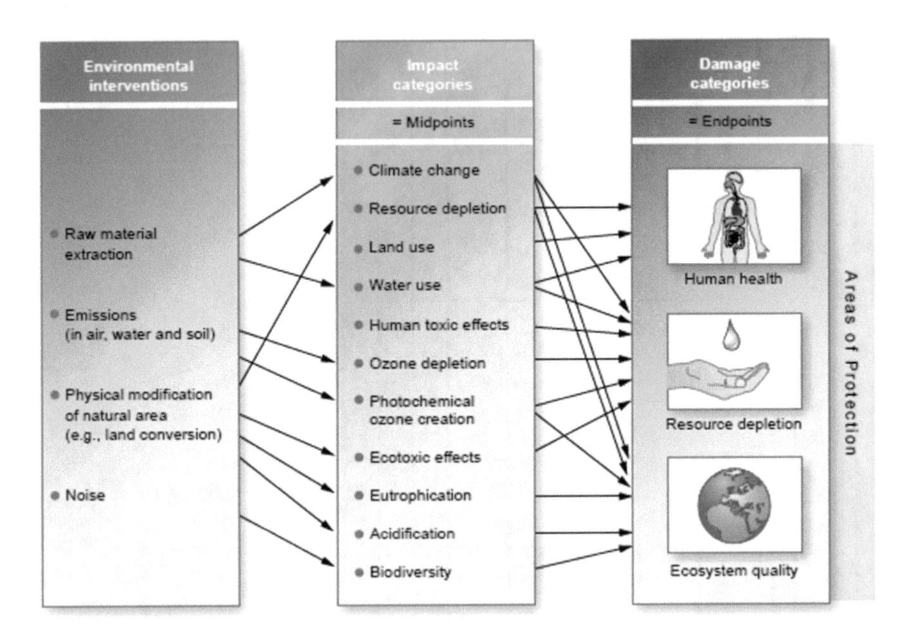

Fig. 2 LCIA framework linking with midpoint and endpoint categories. *Source* Jolliet et al. [31]

Plastics industries, personal care goods, beverages, and automobile sectors have all been known as LCA investment pioneers. Later on, the sectors which started implementing the LCA technique were forestry, mining, oil and gas, the construction/building materials market, the retail and manufacturing industries, and, more significantly, the infrastructure industries such as the energy sector, transportation sector, storage, and communication [29].

3.2 *Life Cycle Costing (LCC)*

The United States Department of Defense (DoD) first used the idea of LCC in the 1960s [16]. Since then various sectors have been involved in determining the optimum expenditure allocation by measuring the costs incurred during the life cycle of a product, service, project, expenditure, and so on [28]. At various life cycle stages, five major cost categories can be used in an LCC review: research, development, and design, primary production, processing, usage, and disposal [28].

LCC was defined by different authors; some of them are summarized in Table 3.

According to the UNEP/SETAC report [64], life cycle costing can be divided into three categories: (i) Conventional LCC; (ii) environmental LCC; and (iii) social LCC.

Table 3 Various LCC definitions

Authors (Year)	Definition
White and Ostwald [69]	"The total cost of an item's life cycle is the amount of all funds spent on it from creation to fabrication to service to the end of its useful life"
Fabrycky and Blanchard [17]	"Every costs related to the product, device, or structure as they are implemented over the product's life cycle"
Rebitzer and Hunkeler [50]	"LCC is an estimate of all costs associated with a product's life cycle that are directly covered by either one or more of the actors in the product life cycle (supplier, manufacturer, user/consumer, EOL-actor (End Of Life-actor), with the addition of externalities that are expected to be internalized in the decision-relevant future"
Code of Practice for Life Cycle Costing [43]	"Life cycle costing is a powerful strategy for assisting managers in making the most cost-effective choices based on alternatives provided to them at various life cycle stages and levels of the life cycle cost estimate"

The conventional life cycle costing is a purely economic evaluation that considers all the direct costs associated with a product. It includes investments related to operational expenses, maintenance of products, and disposal expenses. It is used in many firms and organizations. External costs are often overlooked in this.

Environmental LCC enumerates all costs associated with a product's life cycle that are covered directly by one or more of the individuals involved in the product's life cycle. It also includes the external environmental cost, which includes acidification, eutrophication, climate change, etc., related to different greenhouse gases emission. Both conventional and environmental LCC are included in societal LCC [8].

The Society of Environmental Toxicology and Chemistry (SETAC) has released a code of practice for environmental life cycle costing (LCC), which offers a structure for evaluating decisions as part of product sustainability evaluations using specific, but flexible system boundaries [61]. LCC tries to capture all expenses that are directly met by one or more stakeholders in the product life cycle, which aids in future decision-making. It includes initial cost, installation cost, maintenance cost, downtime, and decommissioning cost (over the lifetime of the project. Currently, there is no standard framework available for performing the LCC of any products or services in the light of sustainability [57].

The different types of costs used in the LCC are:

i. Initial capital cost
ii. Operation and maintenance cost
iii. Disposal/end-of-life cost

i. **Initial Capital Costs**

It is the one-time cost used for the purchase of physical assets, its installation, and commissioning and can be subcategorized as follows:

- Purchase cost includes the cost of land, machinery, refinery development, refinery tank development, non-refinery development, and building construction.
- Acquisitions/finance costs include technical supporting facility cost, equipment transportation cost, and labor insurance.
- Training/installation/commissioning cost is the installation cost of machineries and the training costs for the workers to operate the machinery.

ii. **Operating and Maintenance Costs**

It is the total expenditure involved in the operational production facility. Labor cost, direct resources, direct expenditures, indirect labor, indirect materials, and the cost of the establishment are all part of an asset's operating costs. The estimate of those costs depends on both expected and a real understanding of comparable asset output. Direct labor costs, supplies, fuel power, vehicles, purchased utilities, and operating costs (electricity cost, telecommunication, etc.) are all included in maintenance costs.

iii. **Disposal Cost**

The disposal expense is accrued when an asset is disposed of at the end of its useful life. This cost includes the expense of the asset being destroyed, discarded, or sold, adjusted for any tax deduction or fee upon resale. At the end of its useful life of a product, these costs will be removed from the residual value of the asset.

3.3 Social Life Cycle Assessment

The first attempt was made in the early 1990s to integrate the social dimension into LCA [70]. This helped in developing a methodology that integrates the social aspect of sustainability and addressing the social issues, which include inequality, health, safety, social benefits, corruption, etc.

S-LCA is a systematic approach for assessing the social and socio-economic aspects of products, as well as their possible positive and negative impacts, during their life cycle [72]. The basic idea behind the S-LCA is beyond the environmental impacts; social impacts should also be embodied in process chains so that a similar structure of analysis could be followed. It is one of many types of research that provides an overall view of potential social impacts, which are identified from the complete life cycle of the process or product.

The UNEP/SETAC in 2009, released a Life Cycle Initiative Guidelines for the Social Life Cycle Assessment of Products [64] emphasizing the need for social parameters to be incorporated into LCA. Factors evaluated in S-LCA are the ones that may impact stakeholders directly. Social LCA is defined in the work of UNEP/SETAC [64] as "a systematic process using best available science to collect best available data and report about social impacts (positive and negative) in product life cycles from extraction to final disposal." Goal and scope specification, life cycle inventory analysis, life cycle impact evaluation, and life cycle interpretation are the four main phases of the S-LCA methodology. According to the UNEP/SETAC Guidelines, the social and socio-economic factors assessed in S-LCA can have a positive or negative impact on the company's stakeholders during the product's life cycle.

4 Conclusion

In the present decade, the environmental LCA has progressed to life cycle sustainability assessment (LCSA). LCSA covers all three dimensions of sustainability (environment, economic, and social), whereas basic LCA only considers environmental impacts. Industries are under great pressure from stakeholders to integrate all three dimensions of sustainability. Even though the industries have started implementing environmental aspects, economic and social dimensions are yet to gain importance.

References

1. Abu-Rayash A, Dincer I (2019) Sustainability assessment of energy systems: a novel integrated model. J Clean Prod 212:1098–1116
2. Akber MZ, Thaheem MJ, Arshad H (2017) Life cycle sustainability assessment of electricity generation in Pakistan: policy regime for a sustainable energy mix. Energy Policy 111:111–126
3. Akhtar S, Reza B, Hewage K, Shahriar A, Zargar A, Sadiq R (2015) Life cycle sustainability assessment (LCSA) for selection of sewer pipe materials. Clean Technol Environ Policy 17(4):973–992
4. Balieu R, Chen F, Kringos N (2019) Life cycle sustainability assessment of electrified road systems. Road Mater Pavement Des 20(sup1):S19–S33
5. Baumann H, Tillman A-M (2004) The Hitchhiker's de to LCA. Student Literature
6. Benedetto L, Klemeš J (2009) The environmental performance strategy map: an integrated LCA approach to support the strategic decision-making process. J Clean Prod 17(10):900–906
7. Chen W, Holden NM (2017) Social life cycle assessment of average Irish dairy farm. Int J Life Cycle Assess 22(9):1459–1472
8. Ciroth A, Hildenbrand J, Steen B (2015) Life cycle costing. Sustain Assess Renew Based Prod Methods Case Stud 215–228
9. Contreras-Lisperguer R, Batuecas E, Mayo C, Díaz R, Pérez FJ, Springer C (2018) Sustainability assessment of electricity cogeneration from sugarcane bagasse in Jamaica. J Clean Prod 200:390–401
10. Curran MA (2015) Life cycle assessment: a systems approach to environmental management and sustainability. AIChE. Retrieved from https://www.aiche.org/resources/publications/cep/2015/october/life-cycle-assessment-systems-Ccycleassessment. Int J Life Cycle Assess 23(3):569–580
11. De Luca AI, Falcone G, Stillitano T, Iofrida N, Strano A, Gulisano G (2018) Evaluation of sustainable innovations in olive growing systems: a life cycle sustainability assessment case study in southern Italy. J Clean Prod 171:1187–1202
12. Dillen FJ, Verstraelen LC (eds) (1999) Handbook of differential geometry, vol 7. Elsevier

13. EC (2010) Directive 2010/30/EU of the European Parliament and of the Council of 19 May 2010 on the indication by labelling and standard product information of the consumption of energy and other resources by energy-related products
14. Ekener E, Hansson J, Gustavsson M, Peck P, et al (2016) Integrated assessment of vehicle fuels with life cycle sustainability assessment–tested for two petrol land two biofuel value chains. Report No. 2016:12,f3. The Swedish Knowledge Centre for Renewable Transportation Fuels, Sweden. www.f3centre.se
15. Elhuni RM, Ahmad MM (2017) Key performance indicators for sustainable production evaluation in oil and gas sector. Procedia Manuf 11:718–724. Enterprise development makes a difference? Extr Ind Soc 2:320–327
16. Epstein MJ (1996) Improving environmental management with full environmental cost accounting. Environ Qual Manage 6(1):11–22
17. Fabrycky, Blanchard B (1998) Systems engineering and analysis
18. Finkbeiner M, Schau EM, Lehmann A, Traverso M (2010) Towards life cycle sustainability assessment. Sustainability 2(10):3309–3322https://doi.org/10.3390/su2103309
19. Foolmaun RK, Ramjeawon T (2013) Life cycle sustainability assessments (LCSA) of four disposal scenarios for used polyethylene terephthalate (PET) bottles in Mauritius. Environ Dev Sustain 15(3):783–806
20. Guinée JB, Lindeijer E (eds) (2002) Handbook on lifecycle assessment: operational guide to the ISO standards, vol 7. Springer Science & Business Media
21. Guinée JB, Heijungs R, Huppes G, Zamagni A, Masoni P, Buonamici R, Rydberg T (2011) Lifecycle assessment: past, present, and future. Environ Sci Technol 45(1):90–96.https://doi.org/10.1021/es101316v
22. Guinée J (2016) Life cycle sustainability assessment: what is it and what are its challenges?. In: Taking stock of industrial ecology, Springer, Cham, pp 45–68
23. Halog A, Manik Y (2011) Advancing integrated systems modeling framework for life cycle sustainability assessment. Sustainability 3(2):469–499
24. Hannouf M, Assefa G (2017) Life cycle sustainability assessment for sustainability improvements: a case study of high-density polyethylene production in Alberta, Canada. Sustainability 9(12):2332
25. Heijungs R, Huppes G, Guinée J (2009) A scientific framework for LCA. Deliverable (D15) of work package, 2
26. Hoque N, Biswas W, Mazhar I, Howard I (2019) LCSA framework for assessing sustainability of alternative fuels for transport sector. Chem Eng Trans 72:103–108
27. Horne RE (2009) Limits to labels: the role of eco-labels in the assessment of product sustainability and routes to sustainable consumption. Int J Consum Stud 33(2):175–182
28. Huppes G, van Rooijen M, Kleijn R, Heijungs R, de Koning A, van Oers L (2004) Life cycle costing and the environment. Report of a Project Commissioned by The Ministry of VROM-DGM Leiden, CML
29. Jacquemin L, Pontalier PY, Sablayrolles C (2012) Life cycle assessment (LCA) applied to the process industry: a review. Int J Life Cycle Assess 17(8):1028–1041
30. Janjua SY, Sarker PK, Biswas WK (2020) Development of triple bottom line indicators for life cycle sustainability assessment of residential buildings. J Environ Manag 264:110476
31. Jolliet O, Margni M, Charles R, Humbert S, Payet J, Rebitzer G, Rosenbaum R (2003) IMPACT 2002+: a new life cycle impact assessment methodology. Int J Life Cycle Assess 8(6):324–330
32. Kabayo J, Marques P, Garcia R, Freire F (2019) Life-cycle sustainability assessment of key electricity generation systems in Portugal. Energy 176:131–142
33. Kekäläinen J (2012) Life cycle methods for environmental assessment of nanotechnology
34. Kloepffer W (2008) Life cycle sustainability assessment of products. Int J Life Cycle Assess 13(2):89
35. Kloepffer W (2008) State-of-the-art in life cycle sustainability assessment (LCSA) life cycle sustainability assessment of products. Int J LCA 13(2):89–95. https://doi.org/10.1065/lca2008.02.376

36. Klöpffer W, Grahl B (2014) Life cycle assessment (LCA): a guide to best practice. John Wiley & Sons
37. Krozer J, Vis JC (1998) How to get LCA in the right direction? J Clean Prod 6(1):53–61
38. Li T, Roskilly AP, Wang Y (2018) Life cycle sustainability assessment of grid- connected photo voltaic power generation: a case study of North east England. Appl Energy 3227:465–479
39. Luu LQ, Halog A (2016) Rice husk based bioelectricity vs. coal-fired electricity: life cycle sustainability assessment case study in Vietnam. Procedia CIRP 40:73–78. https://doi.org/10.1016/j.procir.2016.01.058
40. Mahbub N, Oyedun AO, Zhang H, Kumar A, Poganietz WR (2019) A life cycle sustainability assessment (LCSA) of oxymethylene ether as a diesel additive produced from forest biomass. Int J Life Cycle Assess 24(5):881–899
41. Mehmeti A, Pérez-Trujillo JP, Elizalde-Blancas F, Angelis-Dimakis A, McPhail SJ (2018) Exergetic, environmental and economic sustainability assessment of stationary molten carbonate fuel cells. Energy Convers Manag 168:276–287
42. Menikpura SNM, Gheewala SH, Bonnet S (2012) Framework for life cycle sustainability assessment of municipal solid waste management systems with an application to a case study in Thailand. Waste Manage Res 30(7):708–719
43. NATO RTO (2009) Code of practice for life cycle costing (RTO TR-SAS–069, 2009)
44. Nguyen TA, Kuroda K, Otsuka K (2017) Inclusive impact assessment for the sustainability of vegetable oil-based biodiesel—Part I: linkage between inclusive impact index and life cycle sustainability assessment. J Clean Prod 166:1415–1427
45. Oliveira F, Hansson J, Gustavsson M (2015) On environmental LCA for selected transport fuels
46. Onat N, Kucukvar M, Halog A, Cloutier S (2017) Systems thinking for life cycle sustainability assessment: a review of recent developments, applications and future perspectives. Sustainability 9(5):706. https://doi.org/10.3390/su9050706
47. Opher T, Friedler E, Shapira A (2019) Comparative life cycle sustainability assessment of urban water reuse at various centralization scales. Int J Life Cycle Assess 24(7):1319–1332
48. Osorio-Tejada JL, Llera-Sastresa E, Scarpellini S (2017) Liquefied natural gas: could it be a reliable option for road freight transport in the EU? Renew Sustain Energy Rev 71:785–795, Publisher, Lund
49. Pereira CL, Ortega E (2010) Sustainability assessment of large-scale ethanol production from sugarcane. J Cleaner Prod 18(1):77–82
50. Rebitzer G, Hunkeler D (2003) Life cycle costing in LCM: ambitions, opportunities, and limitations. Int J Life Cycle Assess 8(ARTICLE):253–256
51. Rivas Bolivar AJ (2017) Life cycle sustainability analysis (lcsa) of polymer-based piping for plumbing applications
52. Roinioti A, Koroneos C (2019) Integrated life cycle sustainability assessment of the Greek interconnected electricity system. Sustain Energy Technol Assess 32:29–46
53. Ruiz D, SanMiguel G, Corona B, Gaitero A, Domínguez A (2018) Environmental and economic analysis of power generation in a thermophilic biogas plant. Sci Total Environ 633:1418–1428
54. Sala S, Farioli F, Zamagni A (2012) Life cycle sustainability assessment in the context of sustainability science progress (part2), (part2). https://doi.org/10.1007/s11367-012-0509-5
55. Sala S, Farioli F, Zamagni A (2013) Progress in sustainability science: lessons learnt from current methodologies for sustainability assessment: Part1. Int J Life Cycle Assess 18(9):1653–1672
56. Santoyo-Castelazo E, Azapagic A (2014) Sustainability assessment of energy systems: integrating environmental, economic and social aspects. J Clean Prod 80:119–138
57. Schau EM, Traverso M, Lehmann A, Finkbeiner M (2011) Life cycle costing in sustainability assessment—a case study of remanufactured alternators. Sustainability 3(11):2268–2288
58. Settembre-Blundo D,García-Muiña FE, Pini M, Volpi L, Siligardi C, Ferrari AM (2018) Territorial life cycle sustainability assessment (T-LCSA) of Sassuolo Industrial District (Italy)
59. Sharma A, Strezov V (2017) Life cycle environmental and economic impact assessment of alternative transport fuels and power-train technologies. Energy 133:1132–1141

60. Stamford L, Azapagic A (2014) Life cycle environmental impacts of UK shale gas. Appl Energy 134:506–518.
61. Swarr TE, Hunkeler D, Klöpffer W, Pesonen HL, Ciroth A, Brent AC, Pagan R (2011) Environmental life-cycle costing: a code of practice
62. Tye CT (2020) Recent advances in waste cooking oil management and applications for sustainable environment. In: Handbook of research on resource management for pollution and waste treatment, IGI Global, pp 47–63
63. Ugwuoke OS, Oduoza CF (2019) Framework for assessment of oil spill site remediation options in developing countries a life cycle perspective. Procedia Manuf 38:272–281
64. UNEP/SETAC (2009) Guidelines for social life cycle assessment of products, United Nations Environment Program, Paris SETAC Life Cycle Initiative United Nations Environment Programme
65. Valente A, Iribarren D, Dufour J (2017) Life cycle assessment of hydrogen energy systems: a review of methodological choices. Int J Life Cycle Assess 22(3):346–363
66. Vinyes E, Oliver-Solà J, Ugaya C, Rieradevall J, Gasol CM (2013) Application of LCSA to used cooking oil waste management. Int J Life Cycle Assess 18(2):445–455
67. Wang J-J, Jing Y-Y, Zhang C-F, Zhao J-H (2009) Review on multi-criteria decision analysis aid in sustainable energy decision-making. Renew Sustain Energy Rev 13:2263–2278
68. Wang H, Wang L, Shahbazi A (2017) Life cycle assessment of fast pyrolysis of municipal solid waste in North Carolina of USA. J Cleaner Prod 87:511–519
69. White GE, Ostwald PF (1976) Life cycle costing. Manage Account 57(7):39–42
70. Wu R, Yang D, Chen J (2014) Social life cycle assessment revisited. Sustainability 6(7):4200–4226
71. Yıldız-Geyhan E, Yılan G, Altun-Çiftçioğlu GA, Kadırgan MAN (2019) Environmental and social life cycle sustainability assessment of different packaging waste collection systems. Resour, Conserv Recycl, 143:119–132
72. Zamagni A, Pesonen H, Swarr T (2013) From LCA to life cycle sustainability assessment: concept, practice and future directions, 1637–1641. https://doi.org/10.1007/s11367-013-0648-3
73. Zamagni A, Guinée J, Heijungs R, Masoni P, Raggi A (2012) Lights and shadows in consequential LCA. Int J Life Cycle Assess 17(7):904–918
74. Zheng X, Easa SM, Yang Z, Ji T, Jiang Z (2019) Life-cycle sustainability assessment of pavement maintenance alternatives: methodology and case study. J Cleaner Prod 213:659–672
75. Zhou Z, Jiang H, Qin L (2006) Life cycle sustainability assessment of fuels. Fuel 86(1–2):256–263.https://doi.org/10.1016/j.fuel.2006.06.004
76. Zhou HM, Wang TM, Hao WCH et al (2006) Analyses of environmental impact assessment on the development model of circulation economy in Chinese steel industry. In: International Materials Research Conference (Beijing), pp 24–30

A Sustainability Assessment Framework for the Australian Food Industry: Integrating Life Cycle Sustainability Assessment and Circular Economy

Murilo Pagotto, Anthony Halog, Diogo Fleury Azevedo Costa, and Tianchu Lu

Abstract Integrating life cycle techniques and modelling with other approaches to evaluate the eco-efficiency and sustainability of industries and entire supply/value chains has been widely discussed. Life cycle sustainability assessment (LCSA) techniques and modelling capabilities comprehensively evaluate the three sustainability dimensions of complex systems and are able to assess how changes in the system primarily based on circular economy (CE) and other sustainability-based principles will affect the functionality of the system's processes and the effects of those variations in the system as a whole. A systematic literature review was undertaken to analyse the background, the issues and knowledge gaps related to the proposed methodologies as well as the context-specific sustainability aspects faced by the Australian food industry. The systematic review analysed 89 selected studies, and the results demonstrated that sustainability assessment remains a highly complex challenge for the scientific community. Specifically, the development of effective and reliable methods is a nuanced task, particularly when analysing multifaceted systems such as a food supply chain. However, many efforts have been made to extend the focus of the sustainability assessment of industrial systems; but, there is still a lack of approaches that holistically and comprehensively address the triple sustainability dimensions. This chapter draws on peer-reviewed published literature to explore the potential integration of the aforementioned approaches into a holistic and systematic framework for analysing the sustainability of the Australian food industry. The framework assesses environmental, social and economic benefits and effects of implementing sustainable production and consumption processes in the food production system.

M. Pagotto (✉) · A. Halog · T. Lu
School of Earth and Environmental Science, University of Queensland, Brisbane St Lucia, QLD 4072, Australia
e-mail: murilopagotto@live.com

D. F. A. Costa
Institute for Future Farming Systems, Central Queensland University, Rockhampton, QLD 4072, Australia

S. S. Muthu (ed.), *Life Cycle Sustainability Assessment (LCSA)*,
Environmental Footprints and Eco-design of Products and Processes,
https://doi.org/10.1007/978-981-16-4562-4_2

Keywords Life cycle sustainability assessment · Circular economy · Sustainability assessment · Sustainable development · Australian food industry

1 Introduction

Food security is probably the main challenge that food systems worldwide are currently confronting [24, 27]. According to the [24], 'food security is achieved when all people at all times have economic, social and physical access to sufficient, safe, nutritious food that meet their dietary needs as well as their food preferences and allows them to maintain a healthy and active life'. The world population is projected to reach 9.2 billion in 2050 [27]. In this scenario, food production will have to increase at an annual rate of 44 million tones. To meet the projected food demand in 2050, food production will have to increase by approximately 70% [27].

GHG emissions generated by food system are another serious issues created by food systems. If livestock production continues to grow at its current pace, it is expected that by 2050, the sector by itself will exceed the GHG humanity's 'safe operating capacity' [72]. Additionally, any increase in GHG emissions consequently increases the effects of global warming, which is the main driver of climate change globally [13]. Climate change is emerging as a serious issue for food production. It affects the entire food supply chain. Agricultural production, food processing and retail sector are currently facing the effects of global warming and climate change [13]. Rising temperatures directly affect agricultural production areas as well as undesirably affect food processing plants and retail sector [53].

Within food systems, sustainability and sustainable development topics have become important issues in the last and present decades. During the last two decades, several tools and approaches have been developed to assess the sustainability of the food industry and its production processes. However, the majority of these approaches has limitations and does not cover the three dimensions of sustainability [25]. Sustainability is a multifaceted concept and normally the term 'sustainability assessment' is mistakenly used. According to [76], the term is frequently used when just the environmental dimension of sustainability is evaluated, often leading to a reductionist understanding of sustainability.

Governments around the world have been already attempting to reach sustainability and implement sustainable development principles into their policies; nevertheless, global trends and predictions are demonstrating that these efforts are not making enough positive difference [25]. FAO [25] stated that to achieve sustainable development, accurate data and holistic approaches to assess sustainability are necessary. These approaches must have the capacity to efficiently encompass the three dimensions of sustainability: environmental resilience, economic sustainability and social well-being [25, 60].

According to the [83], it is important to predict the effects of shifting environmental and socio-economic policies and production systems of industrial systems. Those changes could create positive and negative effects on the actors involved in the system, so it is recommended to develop scenarios that are capable of incorporating and analysing the sustainability issues that arise during the system's analyses [34, 83]. Researchers involved in sustainability analysis and other disciplines have started to integrate modelling techniques with other approaches to investigate the interactions of industrial systems and their stakeholders [29, 35]. Heairet et al. [35] combined Life Cycle Analysis with Agent-Based Modeling (ABM) to build a framework to analyse the interactions and the system performance of biofuels and bioelectricity systems at the regional level in selected production areas of the United States (US). Florent and Enrico [29] built an approach using Consequential Life Cycle Assessment (CLCA) and ABM to assess the environmental consequences of changes in large-scale transportation policies in Luxemburg. Halog and Bortsie-Ayree [34] emphasized the importance to develop decision support and analytical tools that account for the interactions between the three dimensions of sustainability. Furthermore, to optimize the sustainability of complex industrial systems such as the food industry, the impacts and effects of the decisions made by one actor onto another actor are important to be clearly understood [34].

Within Australia, as around the world, the real nature of the environmental, economic and social impacts created by the food industry is not yet clear [13]. Governments and policy-makers have been reluctant to recognize the seriousness of the impacts of the food systems on the environment as well as the negative socio-economic effects generated by it [13]. It is very important to clearly identify the sustainability of food supply chains and start to measure the impacts created during the production processes involved in it [30].

In recent decades, the Australian government and the governments around the world have started to recognize the links between food production and the environment. Environmental degradation, caused by food production and other industrial systems, generates impacts on food production and food security [67]. Food production strongly relies on ecosystem services to provide natural inputs such as water, natural pest control and climate regulation [67]. Global warming and natural disasters are already affecting some agricultural areas in Australia and other countries, causing financial losses and affecting farmers and isolated communities [85]. Environmental change is already affecting food security, and it is expected that environmental degradation will continue to adversely impact food production worldwide contributing to reducing the quality and affordability of food around the globe, particularly in poor nations and developing countries [27, 85].

1.1 Research Background

The food industry is a major part of Australia's economy. The entire food supply chain, including primary production and manufacturing industries, is valued at more than $ 230 billion [57]. The agricultural sector supplies farm products to the domestic and international markets, particularly the Asian markets. Annually, approximately

77% of Australian agri-food products were exported [57]. Food and beverage manufacturing sectors as downstream industries of agriculture are among the largest industries in Australia [1]. Every year, the Australian food and beverage industry add more than AUD$40 billion to the Australian economy and exports more than AUD$20 billion to its trading partners around the globe [1].

Indeed, the entire food system is economically and socially important to Australia, but it has been causing severe environmental stresses in the process of producing food. It undertakes a capital-intensive activity and requires many inputs, particularly natural resources (e.g. water, land and energy), during its production cycle. It also creates other undesirable outputs such as natural resource depletion, land degradation, greenhouse gas emissions and waste generation, which have adverse consequences in the Australian environment [18].

Agricultural production worldwide is responsible for around 10 per cent of the total anthropogenic GHG emissions [26]. Livestock production activities were estimated to generate approximately 7.1 gigatonnes of CO_2–eq per year; beef and milk production account for around 61 per cent of these emissions [90]. The latter authors emphasize that beef production generates both direct GHG emissions during production and indirect emissions from other activities in its life cycle, such as land clearing, fertilizer production and use and primary energy use during farming of feedstock and transportation of inputs and meat products. Water use, contamination of natural waterways and high consumption of natural resources and primary energy are other environmental issues created during the production of beef and meat products (Gerber et al. 2015) [67].

Greenhouse gas emission and land clearing are amongst the greatest environmental issues in Australia and globally. It threatens the planet's natural functionality affecting ecosystems, food production, biodiversity and climate [67]. The agricultural industry is the largest emitter of methane and nitrous dioxide and the second major source of greenhouse gases (GHG) in Australia [15]. The Australian food-manufacturing sector also significantly contributes to environmental degradation. It produces considerable amounts of undesirable outputs during its production activities [1]. In this complex scenario, changes in the current food production system in Australia must be made to maintain the sustainability of socio-economic and environmental systems [1].

Thus, this chapter has two main objectives. First is to propose an integrated and systematic framework for analysing the sustainability of the Australian food industry. The chapter thus intends to develop the framework that covers and assesses the three dimensions of sustainability of the food system in Australia. Second is to gather the insights of the current state of the methodologies included in it.

2 Materials and Methods

To attain the main aims of this chapter, a SR had been carried out. First, a SR approach was applied to identify studies that considered LCSA and ABM approaches as potential methodologies to measure the sustainability of the food industry in Australia and other countries. Additionally, it identified studies containing information related to the use of CE and Eco-innovation principles to increase the sustainability of industrial systems. We limited the publication period from 2000 onwards (including studies in press) during the studies' retrieving process. Table 1 summarizes the selection criteria for the studies used in the literature review.

The initial scoping performed in the SR focused on the keywords directly related to the topics of this present paper. This approach was selected due to the scope and the number of published articles and other studies containing information related to the subjects. Four electronic databases (Science Direct, Scopus, Google Scholar and Web of Science) were searched using the adopted search strategy. Moreover, the search strategy comprised a search for reports and documents that include relevant information related to the topic of this paper.

Table 1 Selection criteria for LCSA, CE and Australian food systems research studies

Research procedures	i. Selection and assessment of scientific and non-scientific studies related to the sustainability of food systems in Australia and worldwide, LCSA and CE ii. Time period from 2000 onwards iii. Peer-reviewed journal articles and reports from trustworthy sources iv. Assessment of scientific papers cited in selected studies in the previous step v. Backward referencing and author searching vi. Forward referencing and author searching
Databases and other sources	Elsevier, Scopus, Web of Science and Government and private sector websites
Type of analysis	i. Documental survey of peer-reviewed studies published in the assessed databases and sources ii. Analysis of keywords, titles and abstracts iii. Full-text analysis of studies selected in the previous step iv. Analysis of contents and methodologies v. Analysis of cited references vi. Analysis of cited authors
Research bases	i. Searching for terms related to the study (e.g. 'Sustainability Assessment', 'Life Cycle Sustainability Assessment', 'Circular Economy') in titles, keywords and abstracts ii. Scanning of case studies related to the application of the approaches included in the framework iii. Selection of studies according to their publication type iv. Analysis of the distribution and frequency of citation of the selected studies across different journals, sectors and periods

The first stage of the SR focused on the keywords 'Life Cycle Sustainability Assessment' and 'Circular Economy' which are the methodologies and principles included in the proposed sustainability assessment framework. A database search was conducted for each of the keywords (Table 2). The processes consisted of plotting the keywords in the selected databases and performing an electronic search. This process resulted in an unmanageable number of studies (journal articles, reports and books and book chapters). To manage this issue, we excluded some studies using the filtration techniques offered by the databases and applied some of the inclusion and exclusion parameters included in the review's selection criteria. This procedure had decreased the number of studies, but it still resulted in a large number of hits (9947 hits). This issue was managed during the second stage of the systematic review.

The second stage of the SR applied a screening process to analyse the studies retrieved in the first stage of the review. Initially, a title screening of the studies was performed to narrow down the number of results. This approach eliminated a large number of studies that were not considered relevant to this particular research. Duplicated studies were included in the SR results to demonstrate their availability in different databases. The subsequent process involved an analysis of the keywords and a screening of the studies' abstracts selected in the previous processes. The abstract review concentrated in identifying the contents of the studies and evaluating any relevant information that could be useful for this particular paper. These procedures considerably narrowed the results to a manageable number. During this stage, 262 studies (including duplicated studies) were selected for full text review. Prior to the full text review process, a duplicate studies analysis was undertaken resulting in the elimination of 38 studies. Lastly, a full text screening was performed to evaluate the scope and the relevance of the selected. This process assessed the contents of 224 selected studies to define if they are within the scope of this research field. This process reduced the number of studies to 43. The contents and information included in these studies were used to formulate the core of the present review study. Additionally, another 46 peer-reviewed studies containing relevant information and insights were utilized to formulate this review. Figure 1 presents an overview of the review processes performed during the formulation of this particular systematic review.

3 Descriptive Analysis

The descriptive analysis of the final results in this literature review reflects the development of LCSA in the last few years as well as the recent increase in the interest of the application of CE principles. Moreover, it showed how the methodologies analysed in the literature review have been applied in many different fields to assess the environmental and socio-economic impacts of several production processes and industries. The analysis also indicated the diversity of the sustainability assessment and the analysed methodologies in the research field. A total of 33 different journals were represented in the review. Although the analysis illustrated that four journals

Table 2 Search themes and strings for the systematic literature review

Search themes (keywords)	Search strings
Life cycle sustainability assessment	1. (life cycle AND sustainability assessment) AND SUBJAREA (mult OR agri OR bioc OR immu OR neur OR phar OR mult OR ceng OR chem OR comp OR eart OR ener OR engi OR envi OR mate OR math OR phys) AND PUBYEAR > 1999 AND (LIMIT-TO (SUBJAREA,' ENVI')) AND (EXCLUDE (SUBJAREA,' BUSI')) 2. pub-date > 1999 and (life cycle sustainability assessment) AND LIMIT-TO (topics, 'environmental') 3. Timespan: 2000–2020. Indexes: SCI-EXPANDED, SSCI, A&HCI, CPCI-S, CPCI-SSH, BKCI-S, BKCI-SSH, ESCI, CCR-EXPANDED, IC 4. allintitle: 'Life Cycle Sustainability Assessment'
Circular economy	1. (circular economy) AND PUBYEAR > 1999 AND (LIMIT-TO (SUBJAREA,' ENVI')) 2. pub-date > 1999 and TITLE-ABSTR-KEY (Circular Economy) 3. Timespan: 2000–2020. Indexes: SCI-EXPANDED, SSCI, A&HCI, CPCI-S, CPCI-SSH, BKCI-S, BKCI-SSH, ESCI, CCR-EXPANDED, IC 4. Allintitle: 'Circular Economy'

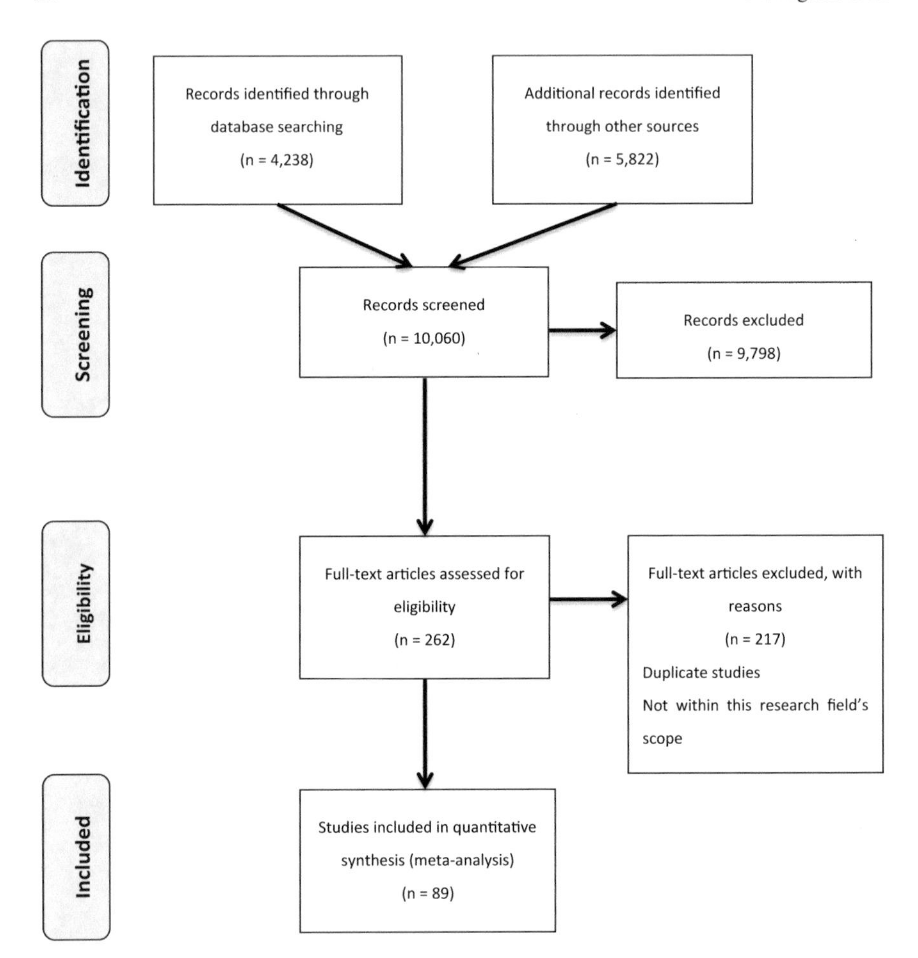

Fig. 1 Overview of the literature review stages and processes including the number of studies analysed during each stage of the literature review

are prominent in the research field as analysed in this paper (International Journal of Life Cycle Assessment; Journal of Cleaner Production; Sustainability; and Ecological Economics). Additionally, the literature base on sustainability assessment of food industry is largely presented in reports produced mainly by international organizations such as FAO and OECD, governments and published books and book chapters produced by researchers and research institutions.

The analysis performed in this research also shows that the number of studies performing sustainability assessment and using the analysed methodologies is growing rapidly in the last two decades. Furthermore, the analysis revealed that only from 2012 onwards, the application of LCSA to measure the sustainability of products started to increase (Table 3). Prior to that period, most of the published studies

Table 3 Distribution of published studies per year of the publication period included in the literature review paper

Year of publication	Number of studies	LCSA studies	CE studies
2000	2	0	2
2001	0	0	0
2002	1	0	0
2003	0	0	0
2004	0	0	0
2005	0	0	0
2006	1	0	0
2007	4	1	2
2008	2	1	1
2009	4	0	2
2010	2	1	0
2011	11	1	0
2012	4	3	1
2013	6	3	3
2014	0	0	0
2015	11	5	4
2016	4	3	1
2017	4	3	1
2018	4	2	2
2019	2	1	1
2020	3	1	2

discussed the development and the issues faced by this particular methodology [88, 89].

Similar conditions as described above have been occurring with regard to the application of CE principles [52, 92]. The application of these principles to increase the sustainability and eco-efficiency of products and production processes are growing fast in the last few years particularly in the European Union, China and Japan [52].

4 Results

The SR results demonstrate the current level of development of the methodologies analysed and included in the proposed sustainability assessment framework in this paper. Additionally, it shows how the methodologies have been used to measure the sustainability of food and industrial systems in the last 16 years. Based on the literature, it appears that LCSA has been considered a useful tool to measure sustainability. However, the methodology has been considered efficient, it still has some issues that need to be addressed. The subsequent sections show the information and results acquired from the studies selected during the SR.

4.1 Life Cycle Sustainability Assessment

To achieve more sustainable production and consumption patterns, life cycle thinking and life cycle-based assessment approaches may play a crucial role in the future. According to [73], the environmental and the socio-economic aspects of the whole product supply chain have to be considered during the life cycle analysis. Furthermore, improving integrated assessment and life cycle analysis methodologies at product development to policy design level is crucial to increase sustainability [73, 82]. To deal with an increasing concern on how to tackle the complex issue of sustainability, several LCSA frameworks have been developed and proposed by different authors. However, not a single study applying the methodology to analyse the sustainability of food production in Australia and other locations was found during the review performed in this chapter.

According to [61], LCSA is one of the most common approaches to evaluate the sustainability of products. Progress and development of LCSA framework are taking the structure of sustainable development to a great extent. The framework is a life cycle-based analysis approach and integrates the most common life cycle analysis methods—environmental LCA, LCC and SLCA. Therefore, it follows the triple bottom line of sustainability assessment methodology [33, 87]. Klöpffer [47] argued that the LCSA model could be an efficient approach in evaluating the 'three pillars' of sustainability. The author also emphasized that to achieve sustainability, the environmental, social and economic aspects of products and production processes have to be assessed against each other. Based on the idea of sustainability analysis, integrated assessment and life cycle perspective analysis, the said author proposed the following LCSA formula:

$$\text{LCSA} = \text{(environmental)LCA} + \text{LCC} + \text{S - LCA}$$

The combination of these three life cycle-based techniques in one integrated assessment framework could provide reliable and robust results during sustainability assessment of industries, production processes and products. This integrated method could also identify the trade-offs between environmental, social and economic dimensions during life cycle analysis [65, 78].

The LCSA methodology aims to produce a detailed representation of the environmental burdens, economic benefits and social impacts created by production systems [82, 87]. It is an efficient approach to evaluate the three pillars of sustainability: environment, economy and society. Onat et al. [64] developed a framework that combines LCSA and multi-criteria decision-making to evaluate the uncertainties related to the implementation of hybrid vehicles in the United States. The framework was used to analyse the sustainability of different hybrid passenger vehicles.

While LCSA is a promising approach to quantify sustainability during life cycle analysis, it also faces some issues [82, 89]. The approach started to gain its momentum in the research community only a few years ago. There are many ongoing discussions about its efficiency as well as about technical requirements and methodological choices. Additionally, selecting sound indicators that cover the three dimensions evaluated by LCSA studies is still a challenge [47]. According to [61], indicators that cover the three dimensions are available and have been used in other studies, but are lacking in completeness and implementation. Niemeijer and Groot [62] also concluded that the major issue is not the lack of LCSA indicators, but the lack of a flawless indicator selection process. Another important factor to be accounted during the LCSA's indicator selection process is the purpose of the analysis that will be undertaken. Heijungs et al. [36] noted that the sustainability analysis of a project, product or production process has a definite interest. Sustainability analysis is commonly used during policy adoption, new technology analysis and implementation as well as product design and commercialization [36]. Indicators are one of the main elements of the analysis, and they are very important during communication and decision-making process [36, 67].

Several organizations have been developing and proposing indicators to be used during LCSA and sustainability analysis as described by Heijungs et al. [36, 89]. These indicators can be effectively used during sustainability analysis. They can be developed for many contexts, and it is very important to carefully select them based on the purpose of the analysis during the indicator selection process [82]. Heijungs, Huppes and Guinée [37] also enforced the importance to take into account life cycle perspective approach during the selection process of indicators for LCSA studies. Numerous authors and organizations [14, 47, 82, 89] argued that sound and effective LCSA indicators should cover the three pillars of sustainability, to include stakeholders' perspectives, environmental impacts and socio-economic aspects.

As briefly mentioned earlier, with regard to the food industry, the literature review performed in this research could not find any study that uses LCSA to assess the sustainability of food systems. The majority of the analysed studies only discuss the methodology itself and the challenges and issues related to its use.

4.2 *Circular Economy Concepts and Principles*

Industrial systems worldwide have been following a linear model of production and consumption of natural resources for decades. These systems require large quantities of non-renewable resources (raw materials and energy) [56]. Recently, private and government organizations started to realize the unseen costs of waste generation as well as the risk exposure to raw material prices fluctuation as imposed by the linear production systems [56]. In this complex scenario, several decision-makers around the globe have agreed that changes in industrial production systems must be made to maintain the sustainability of economic and environmental systems [52]. One possible solution for these issues could be the application of CE principles to increase the sustainability of industrial production and transform the traditional linear system into an efficient circular economy production system.

Circular Economy is defined as an industrial economy that intends to reduce the impacts created by traditional production processes [19]. This model of production relies on renewable energy, diminishes or eliminates the use of toxic chemical compounds and eradicates wastes through eco-redesign of finished products and reuse of by-products through a 'closed loop production' system [19]. Man and Friege [52] stated that the main principle of CE is that the industrial production processes can be designed in a manner that the material flows in the industrial system can be in harmony with natural cycles. Murray et al. [56] described CE as 'an antonym of a linear economy'. According to the authors, to become circular, an economy must attempt to reach the point where it does not create any undesirable effects on the environment. It restores natural systems by ensuring that minimal waste and environmental burdens are produced during the life cycle of products and services. Esa et al. [22] stated that the implementation of CE and closed-loop systems could maximize the use and preservation of natural resources towards sustainable development.

CE policies and principles are based on the idea that strong economy and healthy environment can easily co-exist. Particularly, CE implements strategies to increase the eco-efficiency of industrial systems; therefore, it promotes integration of systems and optimization of services to build an efficient closed-loop system [22, 79]. The main objectives of CE are to optimize production systems applying several principles such as systems thinking, cascade flow of materials and energy, waste minimization and renewable energy use throughout the entire supply chain [19]. The technologies and approaches stated by these principles should be implemented at producers, distributors and final consumers' level to promote the transition of the current production processes and consumption patterns towards more sustainable production and consumption practices [21, 20, 79]. Additionally, CE techniques intend to meet multiple requirements of business models by identifying the key sustainability challenges and opportunities, and then using these to drive changes throughout the company and its value chain, from the business strategy and business model to the operational level [19, 67].

The food industry is a complex system and follows specific patterns [67]. It involves multiple factors and deals with internal and external environmental, social and economic aspects. The implementation of CE principles in the food sector could create environmental and socio-economic benefits and increase its overall sustainability [67]. Moreover, the application of CE principles and techniques into food systems could shape new business strategies that incorporates sustainability throughout all business operations based on life cycle thinking and in cooperation with partners across the value chain. It entails a coordinated set of modifications or novel solutions to products, processes, market approach and organizational structure that leads to a company's enhanced performance and competitiveness [68].

5 The Proposed 'Food Systems Sustainability Assessment Framework' (FSSAF)

5.1 Sustainability Assessment Studies

The importance of incorporating life cycle-based environmental and socio-economic methods in sustainability assessment of food systems has long been recognized by researchers [73]. Life cycle-based methods consider the processes and exchanges that occur throughout the supply chain during food production activities [73]. The use of these methods during sustainability assessment can improve the efficiency of the analysis. The LCSA structure covers all the environmental impacts, costs and externalities, and social impacts created along the supply chain during the production of food products are considered and revealed [70, 28].

Pelletier et al. [71] stated another important factor to consider when undertaking sustainability assessment of food systems—the integration of methods to increase the efficiency of the analysis. Bond et al. [6] argued that combining different approaches to develop a sustainability assessment framework could improve the assessment's results. Additionally, blending computational modelling techniques into it could be an effective approach towards understanding how food systems may respond to changes in policies and production systems (Pelettier et al. 2014). Even though several authors and government institutions fully acknowledge the importance and efficiency of using integrated assessment methods to assess the sustainability of food systems and related products, there is still a lack of broadly accepted and standardized frameworks available [76, 88]. Consequently, the development of transparent, efficient and reproducible frameworks to assess the complexity involved in the sustainability of food systems is necessary.

The notion of sustainability is multidimensional, and its definition and utilization vary significantly with both the context and the user [10]. Indeed, there is a lack of consensus in relation to the use of this term among the scientific community, and increasing complex environmental issues require the enhancement and improvement of the current sustainability assessment methodologies [11]. Nevertheless, this study defines sustainability as the ability of natural and anthropogenic systems to satisfy the needs of present and future generations [84]. Likewise, the meaning of 'sustainability impact assessment' creates discussions among researchers and decision-makers. According to the [63], sustainability impact assessment is an approach that assists decision-makers and researchers to analyse the combined environmental and socio-economic impacts of policies and programmes. Laedre et al. [50] described sustainability impact assessment as a relatively simple tool: the collection and analysis of qualitative and quantitative environmental, economic and social data. In this particular study, sustainability impact assessment was defined as a methodology, which quantifies and evaluates the environmental, economic and social impacts of production systems on the natural environment and society.

The SR reviewed several approaches used to evaluate food systems and products. The application and efficiency of these approaches were evaluated in conjunction with their capability to holistically assess the sustainability aspects of food systems. This factor was considered during the SR because the definition of sustainability is often misinterpreted and most studies do not include the three pillars of sustainability in their assessments. Important issues associated with sustainability assessment remain unresolved: there is a lack of holistic assessment tools for sustainability assessment and the selection of sound indicators continues to be a challenge [50]. As stated by [12], a comprehensive sustainability assessment methodology to assess food systems considering the three pillars of sustainability is currently deficient. Most of the studies analysing the sustainability of food systems only assess environmental impacts using environmental indicators and not considering socio-economic impacts and benefits [12].

The literature search identified several approaches that attempted to evaluate the sustainability of either the entire food industry or selected sub-sectors. Table 4 shows the approaches analysed during the SR. These were selected in accordance to the following criteria:

- The approach was primarily developed or proposed to assess the sustainability of food systems.
- The approach assessed at least one of the three pillars of sustainability.

The results of the literature review revealed constraints in all the sustainability assessment approaches analysed. Most only considered environmental and economic aspects; therefore, the terms sustainability and sustainability impact assessment could be incorrectly used during the development and application of these approaches. For instance, the term was used even when the approach only considered the environmental perspective.

The SR also identified seven approaches that included all three aspects of sustainability assessment in their analyses. López-Ridaura et al. [51] developed a multiscale methodological framework (MMF) for sustainability assessment of new alternatives of natural resource management systems (NRMS) of peasantry systems in Mexico. The framework was designed to evaluate the implications and effects of the implementation of more sustainable NRMS on the environmental, social and economic aspects of small farm operations in the region [51]. Grießhammer et al. [32] proposed a framework called 'Product Sustainability Assessment' (PROSA) to strategically analyse product portfolios, services and products. Additionally, due to its structure, the framework is capable of performing sustainability assessments of product life cycles and supply chains [32]. The Sustainability Assessment of Farming and the Environment (SAFE) framework was proposed by [9] to assess the sustainability of agricultural systems using the three pillars of sustainability of agro-ecosystems [9]. According to the authors:

> The proposed analytical framework is not intended to find a common solution for sustainability in agriculture as a whole, but to serve as an assessment tool for the identification, the development and the evaluation of agricultural production systems, techniques and policies [9], p. 229).

Table 4 Existing assessment methodologies to evaluate the sustainability of food systems

Approach	Level of assessment	Region	Type of assessment	Reference
Agri-LCA	Product	UK	Environmental	[86]
CAPRI	Sector	Europe	Environmental, economic	[38]
DRAM	Sector	Netherlands	Environmental, economic	[38]
FARMIS	Sector	Germany	Environmental, economic	[74]
MMF	Field/farm	Mali	Environmental, economic and social	[51]
MODAM	Sector	Germany	Environmental, economic	[75]
PASMA	Sector	Austria	Environmental, economic	[77]
PROMAPA.G	Sector	Spain	Environmental, economic	[4]
PROSA	Product	Germany	Environmental, economic and social	[32]
RAUMIS	Sector	Germany	Environmental, economic	[44]
SAFE	Farm	Global	Environmental, economic and social	[9]
SALCA	Product	Switzerland	Environmental	[58]
LCSA	Product	Global	Environmental, economic and social	[82]
Slow Food Presidia	Product	Italy	Environmental, economic and social	[69]
SUSFANS	Sector	European Union	Environmental, economic and social	[91]
ASLCA	Sector	New Zealand	Environmental and economic	[11]
LCA4CSA	Sector	Colombia	Environmental	[2]
SustainFARM	Sector	European Union	Environmental, economic and social	[55]

Peano et al. [69] proposed a multi-criteria approach called Slow Food Presidia to assess the sustainability of agri-food systems. The authors selected environmental, economic, social and cultural indicators from existing methodologies. When reliable indicators were not available, suitable indicators were designed using consultation with experts and stakeholders involved in food production. Recently, [91] developed an integrated sustainability assessment approach to evaluate the sustainability of food systems in the European Union. The approach was designed to assess, identify and communicate the current state of the sustainability of agricultural and food systems in

Europe. Lastly, [55] designed a Delphi-style approach to investigate the sustainability of sustainable farming systems in Europe.

The proposed framework in this paper provides a guideline in identifying and evaluating accurately sustainability issues and challenges faced by the food industry in Australia. It demonstrates the process of determining the inefficient processes that are currently used during food production in Australia using an integrated approach (Fig. 4). Once the inefficient processes are recognized though the LCSA analysis, the framework can focus on identifying the ways of implementing new technologies and measures to improve the industry's sustainability. This attempt will also use modelling approaches to detect and assess the accompanying benefits and impacts of the proposed changes that will affect the industry itself, its stakeholders and society more generally.

In recent decades, there has been rising concern among governments and private sector leaders about sustainability issues (e.g. climate change, workers welfare and resource constraints), which have a significant impact on businesses and society [7, 16]. Thus, a novel approach to address sustainability-related challenges is needed in addition to opportunities for growth, cost reduction, competitive advantages and the promotion of well-being [7].

LCSA is an approach that intends to meet these multiple requirements by identifying the key sustainability challenges facing industrial systems. This identification is important to uncover opportunities to drive change throughout businesses and their value chains, from the business strategy and model to the daily operations [18, 22]. The LCSA methodology can also be used to evaluate the development and application of more sustainable production systems that incorporate sustainability throughout all operations based on life cycle thinking and in cooperation with partners across the value chain [31, 65].

The evaluation of the sustainability of industrial systems is important to understand how these systems affect the environment and human society. Such assessments are nuanced tasks, and several LCSA frameworks have been developed and proposed by different authors for this purpose. The use of LCSA methodology has also been considered worldwide during policy and technology design to increase the sustainability of production systems [21]. The approach could also be useful to evaluate how the development of sustainable products (goods and services) and processes can increase the overall sustainability of systems when compared with traditional practices [21, 49, 80].

Despite recent developments in the LCSA methodology, not a single study applying the methodology to analyse the sustainability of food production in Australia and other locations was found during the review performed to produce this chapter. Table 5 shows some of the recent studies using LCSA for several production systems and technologies in different countries. Onat et al. [65] also assessed the current development of the LCSA approach through a literature review and found a growing interest in the use of this method in the last decade; however, its application has been limited mainly to environmental sciences. Indeed, from 2000 to 2017, only 56 studies quantitatively applied the LCSA to analyse the sustainability of products and industrial systems [65].

Table 5 Recent studies applying the LCSA methodology

Authors (Year)	Scope	Country or region	Level of assessment
Atilgan and Azapagic [5]	LCSA of the Turkish electricity sector	Turkey	Environmental: Resource depletion, climate change, emissions Economic: Costs Social: Provision of employment, worker safety, energy security
Hossaini et al. [40]	Environmental and socio-economic impact assessment of construction and buildings	Canada	Environmental: Resource use, climate change, impacts on air, water and soil Economic: Costs Social: Safety, affordability
Akhtar et al. [3]	Selection of sewer pipe material application based on the results of two different LCSA approaches	North America	Environmental: Resource depletion, climate change, emissions Economic: Production and environmental costs
Huang and Mauerhofer [42]	Sustainability assessment of ground source heat pump using LCSA techniques	China	Environmental: Resource use, climate change, impacts on air, water and soil Economic: Costs Social: Provision of employment
Onat et al. [64]	Framework combining LCSA and multi-criteria decision-making to evaluate the uncertainties related to the implementation of hybrid vehicles in the United States	United States	Environmental: Resource use, climate change, impacts on air, water and soil Economic: Value added, imports Social: Employment, injuries, income, government taxes
Hossaini et al. [40]	Integration of regional characteristics and LCSA to assist in sustainability design and decision-making for net-zero buildings	Canada	Environmental: Resource use, climate change Economic: Overall costs of new technologies Social: Social benefits

(continued)

Table 5 (continued)

Authors (Year)	Scope	Country or region	Level of assessment
Onat et al. [64]	Integration of LCSA and system dynamic approaches to compare different hybrid vehicles and to build scenarios to test their long-term sustainability in the United States	United States	Environmental: Climate change, emissions Economic: Contribution to gross domestic product, ownership costs Social: Employment, human health
Yu and Halog [87]	LCSA framework to evaluate the sustainability of solar photovoltaic systems in Australia	Australia	Environmental: Climate change, emissions and resource use Economic: Production and installation costs Social: Employment, health and safety, contribution to society

5.2 *The Proposed Framework*

The food industry is a complex system and follows specific patterns that involve multiple factors and deals with both internal and external environmental, social and economic aspects [67]. Addressing the research gaps in the evaluation of the sustainability of food systems is important to maintain the balance of food supply and production without creating detrimental environmental and socio-economic impacts [65]. Additionally, holistic assessment of the food sector could create environmental and socio-economic benefits and increase its overall sustainability [45]. The integration of life cycle techniques and modelling with other tools to evaluate the eco-efficiency and sustainability of industries and entire supply/value chains have been widely discussed. LCSA evaluates the three sustainability dimensions of complex systems. This approach can also be used to evaluate how changes in the system will affect its overall sustainability.

According to [45], it is important to analyse the background, issues and knowledge gaps related to sustainability assessment methodologies. Specifically, the development of effective and reliable methods is a nuanced task, particularly when analysing multifaceted systems such as a food supply chain. Although there has been significant effort to extend the focus of sustainability assessments of industrial systems, there is a lack of approaches that holistically and comprehensively address the triple sustainability dimensions [83].

Sustainability assessment has presented a complex challenge for the scientific community. Specifically, the development of effective and reliable methods is a difficult task, particularly when analysing complex systems such as the food supply chain [60, 76]. Ness et al. [60] presented a framework to categorize the spectrum of available sustainability assessment tools, and demonstrated that most of the current well-established tools are not fully capable of completely assessing the sustainability of products and industrial systems. Thus far, most methods to assess sustainability only consider one of the three dimensions. With regard to the food industry, and particularly at its primary production stage (agriculture), there is a lack of reference tools to assess sustainability [25, 46, 54, 64]. Supranational bodies, such as the European Union [20], have noted that better technical knowledge on the environmental, economic and social impacts of food production is needed.

Currently, there is no standardized framework to analyse sustainability of adaptive complex systems. Furthermore, there is no study that extends the traditional LCA to account the social and economic pillars of sustainability in the Australian food industry [47, 61]. The proposed sustainability assessment framework in this paper is expected to clarify important questions and concerns related to the eco-efficiency, sustainability and competitiveness of the Australian food industry. Moreover, it includes new approaches and methodologies aiming to efficiently evaluate the consumption and production patterns of the Australian food system as well as to construct a sound and comprehensive database containing data related to environmental impacts and costs, resource efficiency, economic and social importance of the Australian food industry.

To minimize the environmental and socio-economic impacts as well as increase sustainability, food industry stakeholders must develop an evaluation methodology, a set of sustainability criteria and a sound technical guide that cover the identification of more sustainable food production processes or improvement of existing procedures [20]. Schader et al. [76] evaluated several sustainability assessment tools commonly applied to food systems with respect to their range and precision. The study demonstrated that all the approaches analysed have limitations in their scope. Additionally, not one among the evaluated approaches completely covers the three dimensions of sustainability, although researchers have long recognized the importance of incorporating life cycle-based environmental and socio-economic methods in the sustainability assessment of food systems [73]. Life cycle-based methods consider the processes and exchanges throughout the supply chain during food production activities [70], and their use during sustainability assessment can improve the efficiency of the analysis. The LCSA structure considers the environmental impacts, costs and externalities and the social consequences of the food production supply chain [47, 70, 28].

Although several authors and government institutions fully acknowledge the importance and efficiency of using integrated methods to assess the sustainability of food systems and related products, there is still a lack of broadly accepted and standardized frameworks [76, 88]. Consequently, the development of transparent, efficient and reproducible frameworks to assess the complexity of the sustainability of food systems is necessary.

The proposed framework in this research provides a guideline for identifying and evaluating sustainability challenges facing the food industry in Australia. The framework is named the 'Food Systems Sustainability Assessment Framework' (FSSAF) and aids to identify the inefficient processes that are currently used during food production in Australia using an integrated LCSA approach (Fig. 2). Once

LCSA Inventory

Environmental		Economic		Social	
Inputs	**Outputs**	**Inputs**	**Outputs**	**Inputs**	**Outputs**
Energy Water Transportation	GHG Wastewater General	Production costs Waste	Income generation Profits	Workforce	Social benefits Social

LCSA Impact

Environmental Impact Categories
Acidification potential: kg SO_2 eq.
Climate Change: GWP 100; kg CO_2 eq.
Depletion of abiotic resources: kg antimony eq.
Depletion of abiotic resources: fossil fuels; MJ
Eutrophication: kg PO_4 eq.
Freshwater aquatic ecotoxicity: kg 1,4-dichlorobenzene eq.
Human ecotoxicity: kg 1,4-dichlorobenzene eq.
Marine aquatic ecotoxicity: kg 1,4-dichlorobenzene eq.
Ozone depletion: kg CFC-11 eq.
Photochemical oxidation: kg ethylene eq.
Terrestrial ecotoxicity: kg 1,4-dichlorobenzene eq.

Cost Impact Categories
Energy: AUD
Labour: AUD
Other inputs: AUD
Tax, depreciation, rates and other overheads: AUD
Transportation: AUD
Water: AUD
Waste management: AUD

Costs Impact Categories
Child Labour: CH medium risk
Community Engagement: CE medium risk
Equal Opportunity: EO medium risk
Fair Salary: FS medium risk
Freedom of Association and Bargain: FB medium risk
Gender Equality: GE medium risk
Healthy and Safety: HS medium risk
Indigenous Rights: IR medium risk
Injuries and Fatalities at Work: IF medium risk
Local Community: LC medium risk
Local Employment: LE medium risk
Working Conditions: WC medium risk
Working Hours: WH medium risk

Fig. 2 The proposed FSSAF to assess the sustainability of the Australian food industry sectors. *Data Source* OpenLCA [66]

such processes are recognized, the focus will be on identifying methods for implementing new technologies and measures to improve the industry's sustainability. This attempt also used LCSA modelling approaches to detect and assess the accompanying benefits and impacts of the proposed changes for the industry, its stakeholders and society.

The conceptual FSSAF framework set out in this study attempts to develop a new approach integrating LCSA and CE to efficiently evaluate the resource efficiency and consumption and production patterns in the Australian food system. The framework integrates different approaches for quantifying the impacts (positive and negative) of industrial systems, and identifies and evaluates sustainable production processes and their potential benefits for the industry and its stakeholders. This innovative approach intends to address the industry's priority goals while minimizing its environmental, economic and social impacts. Further, it intends to assess current and future challenges that could affect the industry's functionality. The framework is expected to support the Australian food industry's stakeholders in evaluating current resource and production efficiency and waste management processes. Finally, stakeholders and decision-makers can use the framework to perform holistic analysis covering the social, environmental and economic dimensions of the food industry to support sound policy recommendations.

The FSSAF addresses three critical concerns regarding the Australian food production system. First, it evaluates and quantifies the environmental burdens and socio-economic impacts of the Australian food industry using LCSA. Second, it identifies and assesses sustainable food production technologies and approaches applying the concepts CE. Lastly, it assesses the impacts and benefits of implementation of the recommended approaches in the Australian food production system using the developed LCSA framework for sustainability assessment. The framework aims to produce a detailed representation of the environmental burdens, economic benefits and social impacts of food production in Australia [82].

The LCSA methodology incorporated in the FSSAF includes an attributional life cycle assessment to evaluate the environmental impacts created during the production of food products in Australia. This approach quantifies the environmental impacts, resource use and waste involved in the production of a particular food product [39]. The framework also examines the product's life cycle using an LCC method to deliver an economic analysis of the processes involved [43]. Finally, SLCA is used to analyse the social impacts (both positive and negative). This technique was recently developed to assess the social and socio-economic aspects of products and their life cycles [81]. According to [81], SLCA can be an effective complimentary approach to provide additional social and socio-economic data during LCA of products.

The FSSAF also aims to evaluate how the implementation of novel technologies and approaches can increase the sustainability, feasibility and competitiveness of the Australian food industry. These approaches are proposed to improve problematic aspects of the current technologies identified during the LCSA analysis. CE concepts and principles guided adjustments to the approaches to produce the desired improvements and their results. The implementation of these concepts could reduce the impact of traditional production processes and thus, is vital. Finally, applying CE principles in the Australian food system could increase its competitiveness, decrease costs and increase the value of current production systems [23].

There are currently no standardized frameworks to analyse the sustainability of adaptive complex systems such as the food industry. No study has yet extended the traditional LCA to consider the social and economic pillars of sustainability in the Australian food industry [47, 61]. The proposed sustainability assessment framework in this study is expected to clarify important questions and concerns related to the sustainability and competitiveness of the Australian food system. Moreover, it includes approaches and methodologies to efficiently evaluate the consumption and production patterns of the system, and to construct a sound and comprehensive database of the environmental impacts and costs, resource efficiency and economic and social importance of the Australian food industry.

Lastly, the FSSAF attempts to demonstrate the benefits of shifting the food production system in Australia into a sustainable system based on the principles of CE. It uses the LCSA methodology to build models of food production processes to evaluate the effects of implementing sustainable technologies (Fig. 3). Another benefit of the proposed framework is that models and different approaches can be included to improve its usability and efficiency, when required.

Overall, the approach proposed in this study expects to spur initiatives to develop a sound and comprehensive database that covers the environmental, economic and social aspects of the Australian food industry. While it could use existing LCSA studies and guidelines, the study is equally concerned with detecting the main issues facing the current structure to improve it. Additionally, the research underscores the benefits of transforming Australian food production into a sustainable system based on the principles of CE.

The implementation of CE could increase the sustainability and competitiveness of the food industry in Australia and worldwide. These principles promote

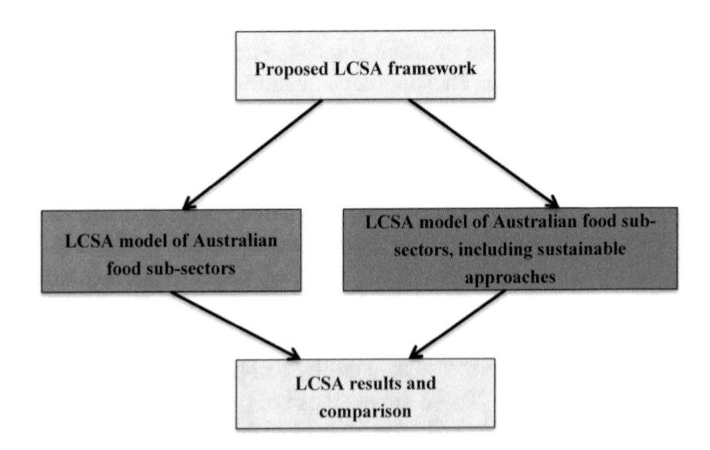

Fig. 3 How the FSSAF methodology evaluates improvements in the sustainability of the Australian food industry

'closed-loop' production systems that aim to reduce resource consumption and waste generation [17]. The main objective is to optimize production systems by applying several principles (including system thinking, cascade flow of materials and energy, waste generation minimization and renewable energy sources) throughout the entire supply chain [17]. The approaches stated by these principles should be implemented at all production levels to promote the transition of the current processes towards sustainability [18]. The implementation of CE and eco-innovation principles in the Australian food system could solve certain current sustainability issues. For example, it could mend inefficiencies along the food supply chain (e.g. loss of productivity, energy and natural resources; waste generation and associated costs) and improve the industry's economic and social dimensions by generating new market opportunities for sustainable products and green jobs in rural and urban communities.

6 Conclusions and Further Studies

6.1 Conclusions

In the last few decades, natural resource depletion, environmental issues and population growth are exerting additional pressure on food industry in Australia and globally. International competitiveness is increasing, which directly affects the domestic food industry in Australia, particularly due to cheaper imported food products. Based on these circumstances, maintaining food security in Australia will be a major challenge in the years to come. The actual food production system in Australia consumes large amounts of resources during its production processes. Under business-as-usual scenarios, this system is on course to create irreversible environmental impacts in the Australian environment which risks jeopardizing future food production in some important food basket regions in the country. Additionally, rapid depletion of natural resources and energy sources will raise the production costs (raw materials, energy and transportation) of food in Australia. Based on these facts, a comprehensive sustainability assessment of the entire food supply chain system in Australia is necessary to construct an efficient model that will guarantee a sufficient and nutritional food supply for its inhabitants in the near future.

Accurately quantifying the resource efficiency of the Australian food industry is extremely important for predicting major environmental issues as well as for implementing measures to prevent them in the future. The conceptual framework set out in this paper attempts to develop a new approach combining different methodologies (LCSA and CE) to efficiently evaluate the resource efficiency and consumption and production patterns of the Australian food system. This innovative approach intends to address the industry's priority goals while minimizing its environmental, economic and social impacts. Furthermore, it intends to assess the current and future issues and challenges that could affect the industry's functionality. The framework is expected to support the Australian food industry's stakeholders in evaluating resource and production efficiency and waste management processes currently used by the industry. It is projected to identify and evaluate the possible benefits and impacts that the implementation of new technologies and approaches will create in the industry and to its stakeholders. Finally, the stakeholders involved in food production in Australia and decision-makers can use the framework to perform holistic analysis covering the social, environmental and economic dimensions of the food industry for sound policy recommendations.

6.2 Further Studies

The importance of evaluating the sustainability of production systems is well proven. Therefore, further studies should focus on identifying better approaches and to measure the environmental and socio-economic impacts of food and other production systems. Additionally, application of the FSSAF in a real scenario would serve to test its effectiveness and identify theoretical, methodological and practical issues that could be corrected in further studies.

The approach developed in this research could be tested to verify its effectiveness in decision-making processes and policy design and implementation. As this research only considered application of the framework to evaluate food system sustainability, an investigation of its capability to support the development of sustainable policies and regulations should be considered. The FSSAF proposed in this project could be used to verify the efficiency and progress of programmes proposed and implemented by the Australian government and other institutions. For example, the framework can be used to evaluate the effects generated by the implementation of the Queensland Biofutures Action Plan in the overall sustainability of production systems and rural communities in Queensland.

With regard to the implementation of sustainable production processes and CE principles to increase the sustainability of food systems, this research can be further extended and developed to solve some limitations and challenges faced during the design, implementation and monitoring of policies including these principles. According to Korhonen et al. [48], CE approaches to scientific and research basis are still in its early development, and several areas need more scientific research-based studies. The use of biomaterials and biofuels to increase systems' circularity are one of the main concepts of CE. However, the utilization of these sustainable materials are increasing in recent times. Korhonen et al. [48] pointed out that the assessment of the environmental impacts created by their production and use still faces several unresolved methodological and theoretical limitations. Further studies to improve the sustainability assessment framework and the methodological and theoretical approaches proposed and presented in this research project could be useful to answer some questions and solve some of the issues and challenges mentioned in this paragraph.

References

1. AFGC (2019) State of the industry report. Australian Food & Grocery Council, Canberra
2. Acosta-Alba I, Chia E, Andrieu N (2019) The LCA4CSA framework: using life cycle assessment to strengthen environmental sustainability analysis of climate smart agriculture options at farm and crop system levels. Agric Syst 171:155–170
3. Akhtar S, Reza B, Hewage K, Shahriar A, Zargar A, Sadiq R (2018) Life cycle sustainability assessment (LCSA) for selection of sewer pipe materials. Clean Technol Environ Policy 17(4)
4. Asensio L, Barreda RGd, Ruiz M, Diego J-LMd, Miqueleiz E (2011) An application of a positive mathematical programming model to analyse the impact of agricultural policy measures in the Spanish agricultural sector. In: Ad Prado, AJB Luiz, HC Filho (eds), Computational methods for agricultural research: advances and applications. IGI Global, Hershey, pp 175–98
5. Atilgan B, Azapagic A (2016) An integrated life cycle sustainability assessment of electricity generation inTurkey. Energy Policy 93:168–186
6. Bond R, Curran J, Kirkpatrick C, Lee N, Francis P (2001) Integrated impact assessment for sustainable development: a case study approach. World Dev 29(6):1011–1024
7. Bossle MB, Barcellos MDD, Vieira LM (2015) Eco-innovative food in Brazil: perceptions from producers and consumers. Agric Food Econ 3(8):1–18
8. Cao K, Feng X, Wan H (2009) Applying agent-based modeling to the evolution of eco-industrial systems. Ecol Econ 68(11):2868–2876
9. Cauwenbergh NV, Biala K, Bielders C, Brouckaert V, Franchois L, Cidad VG, Hermy M, Mathijs E, Muys B, Reijnders J, Sauvenier X, Valckx J, Vanclooster M, Veken BVd, Wauters E, Peeters A (2007) SAFE—a hierarchical framework for assessing the sustainability of agricultural systems. Agr Ecosyst Environ 120(2–4):229–242
10. Chandrakumar C, McLaren SJ (2018) Towards a comprehensive absolute sustainability assessment method for effective Earth system governance: Defining key environmental indicators using an enhanced-DPSIR framework. Ecol Ind 90:577–583
11. Chandrakumar C, McLaren SJ, Jayamaha NP, Ramilan T (2018) Absolute sustainability-based life cycle assessment (ASLCA): a benchmarking approach to operate agri-food systems within the 2 degrees C global carbon budget. J Ind Ecol 23(4):906–917
12. Chaudhary A, Gustafson D, Mathys A (2018) Multi-indicator sustainability assessment of global food systems. Nat Commun 9(1):1–13
13. Chen W, Holden NM (2018) Tiered life cycle sustainability assessment applied to a grazing dairy farm. J Clean Prod 172:1169–1179
14. Cinelli M, Coles SR, Jørgensen A, Zamagni A, Fernando C, Kirwan K (2013) Workshop on life cycle sustainability assessment: the state of the art and research needs—November 26, 2012 Copenhagen, Denmark. Int J Life Cycle Assess 18:1421–1424
15. DISER (2020) Australia's Emissions Projections 2020. Australian Government Department of Industry, Science, Energy and Resources, Canberra
16. Demirel P, Kesidou E (2011) Stimulating different types of eco-innovation in the UK: government policies and firm motivations. Ecol Econ 70(8):1546–1557
17. EMF (2013) Towards the circular economy: economic and business rationale for an accelerated transition. Ellen MacArthur Foundation, London
18. EMF (2015) Towards the circular economy—economic and business rationale for an accelerated transition. Ellen MacArthur Foundation, London
19. EMF (2018) Cities and the circular economy for food. Ellen MacArtur Foundation, Cowes
20. EU (2014) How can we move towards a more resource efficient and sustainable food system. European Union, Brussels
21. Ekener E, Hansson J, Larsson A, Pecke P (2018) Developing life cycle sustainability assessment methodology by applying values-based sustainability weighting—tested on biomass based and fossil transportation fuels. J Clean Prod 181:337–351
22. Esa MR, Halog A, Rigamonti L (2016) Developing strategies for managing construction and demolition wastes in Malaysia based on the concept of circular economy. J Mater Cycles Waste Manag.

23. Esty DC, Porter ME (1998) Industrial ecology and competitiveness. Yale Law School, New Haven
24. FAO (1996) Report of the world food summit-Part I. Food and Agriculture Organization of the United Nations, Rome
25. FAO (2014) SAFA guidelines: sustainability assessment of food and agriculture systems. Food and Agriculture Organization of the United Nations, Rome
26. FAO (2017) The future of food and agriculture—trends and challenges. Food and Agriculutre Organization of the United Nations, Rome
27. FAO (2020) The state of food security and nutrition in the world 2020: transforming food systems for affordable healthy diets. Food and Agriculture Organization of the United Nations, Rome
28. Finkbeiner M, Lehmann A, Schau EM, Traverso M (2010) Towards life cycle sustainability assessment. Sustain 2(10):3309–3322
29. Florent Q, Enrico B (2015) Combining agent-based modeling and life cycle assessment for the evaluation of mobility policies. Environ Sci Technol 49(3):1744–1751
30. Garofalo P, D'Andrea L, Tomaiuolo M, Venezia A, Castrignano A (2017) Environmental sustainability of agri-food supply chains in Italy: the case of the whole-peeled tomato production under life cycle assessment methodology. J Food Eng 200:1–12
31. Gloria T, Guinée J, Kua HW, Singh B, Lifset R (2017) Charting the future of life cycle sustainability assessment: a special issue. J Ind Ecol 21(6):1449–1453
32. Grießhammer R, Buchert M, Gensch C-O, Hochfeld C, Manhart A, Rüdenauer I (2007) PROSA—Product Sustainability Assessment Institute of Applied Ecology, Freiburg
33. Halog A, Manik Y (2011) Advancing integrated systems modelling framework for life cycle sustainability assessment. Sustainability 3:469–499
34. Halog A, Bortsie-Ayree NA (2013) The Need for Integrated Life Cycle Sustainability Analysis of Biofuel Supply Chains. In: Z Fang (ed) Biofuels—economy, environment and sustainability. https://doi.org/10.5772/52700
35. Heairet A, Choudhary S, Miller SA, Xu, M (2012) Beyond life cycle analysis: Using an agent based approach to model the emerging bioenergy industry. In: International Symposium on Sustainable Systems and Technology, Boston
36. Heijungs R, Huppes G, Guinee JB (2010) Life cycle assessment and sustainability analysis of products, materials and technologies Toward a scientific framework for sustainability life cycle analysis. Polym Degrad Stab 95:422–428
37. Heijungs R, Settanni E, Guinée J (2013) Toward a computational structure for life cycle sustainability analysis: unifying LCA and LCC. Int J Life Cycle Assess 18(9):1722–1733
38. Helming JFM (2005) A model of Dutch agriculture based on Positive Mathematical Programming with regional and environmental applications, PhD thesis, Wageningen University
39. Horne R, Grant T, Verghese K (2009) Life cycle assessment: principles, practice and prospects. CSIRO Publishing, Collingwood
40. Hossaini N, Hewage K, Sadiq R (2015) Spatial life cycle sustainability assessment: a conceptual framework for net-zero buildings. Clean Technol Environ Policy 17:2243–2253
41. Hossaini N, Reza B, Akhtar S, Sadiq R, Hewage K (2015) AHP based life cycle sustainability assessment (LCSA) framework: a case study of six storey wood frame and concrete frame buildings in Vancouver. J Environ Planning Manage 58(7):1217–1241
42. Huang B, Mauerhofer V (2016) Life cycle sustainability assessment of ground source heat pump in Shanghai China. J Clean Prod 119:207–214
43. Hunkeler D, Lichtenvort K, Rebitzer G (eds) (2008) Environmental life cycle costing. Society of Environmental Toxicology and Chemistry, New York
44. Julius C, Moller C, Osterburg B, Sieber S (2003) Indicatorsfor a sustainable land use in the "regionalised agricultural and environmental information system for Germany." Agrarwirtschaft 52(4):184–194
45. Jurgilevich A, Birge T, Kentala J, Korhonen-Kurki K, Saikku JP, Schösler H (2016) Transition towards circular economy in the food system. Sustainability 8(1):69

46. Karvonen J, Halder P, Kangas J, Leskinen P (2017) Indicators and tools for assessing sustainability impacts of the forest bioeconomy. For Ecosyst 4(1):2
47. Klöpffer W (2008) Life cycle sustainability assessment of products (with comments by Helias A. Udo de Haes, p. 95). Life Cycle Sustain Assess Prod 13(2):89–95
48. Korhonen J, Honkasalo A, Seppälä J (2018) Circular economy: the concept and its limitations. Ecol Econ 143:37–46
49. Lacy P, Rutqvist J (2015) Waste to wealth: creating advantage in a circular economy. Palgrave Macmillan, Hampshire
50. Laedre O, Haavaldsen T, Bohne RA, Kallaos J, Lohne J (2014) Determining sustainability impact assessment indicators. Impact Assess Proj Apprais 33(2):1–10
51. López-Ridaura S, Keulen Hv, Ittersum MKv, Leffelaar PA (2005) Multi-scale sustainability evaluation of natural resource management systems: Quantifying indicators for different scales of analysis and their trade-offs using linear programming. Int J Sustain Dev & World Ecol 12(2):81–97
52. Man Rd, Friege H (2016) Circular economy: European policy on shaky ground. Waste Manage Res 34(2):93–95
53. Moon YB (2015) Simulation modeling for sustainability: a review of the literature. Syracuse University, Syracuse
54. Morawicki RO (2012) Handbook of sustainability for the food sciences. Wiley, Hoboken
55. Mullender SM, Sandor M, Pisanelli A, Kozyra J, Borek R, Ghaley BB, Gliga A, von Oppenkowski M, Roesler T, Salkanovic E, Smith J, Smith LG (2020) A Delphi-style approach for developing an integrated food/non-food system sustainability assessment tool. Environ Impact Assess Rev 84
56. Murray A, Skene K, Haynes K (2017) The circular economy: an interdisciplinary exploration of the concept and application in a global context. J Bus Ethics 140(3):369–380
57. NFF (2017) Food, fibre & forestry facts: a summary of Australia's agriculture sector. National Farmers Federation, Barton
58. Nemecek T, Huguenin-Elie O, Dubois D, Gaillard G, Schaller B, Chervet A (2011) Life cycle assessment of Swiss farming systems: II. Extensive and intensive production. Agric Syst 104:233–245
59. Nemeck T, Kagi T (2007) Life cycle inventories of agricultural production systems. Agroscope Reckenholz, Zurich
60. Ness B, Urbel-Piirsalua E, Anderbergd S, Olssona L (2007) Categorising tools for sustainability assessment. Ecol Econ 60:498–508
61. Neugebauer S, Martinez-Blanco J, Scheumann R, Finkbeiner M (2015) Enhancing the practical implementation of life cycle sustainability assessment e proposal of a Tiered approach. J Clean Prod 102:498–508
62. Niemeijer D, Groot RSd (2008) A conceptual framework for selecting environmental indicator sets. Ecol Ind 8(1):14–25
63. OECD (2010), Guidance on sustainability impact assessment
64. Onat NC, Kucukvar M, Tatari O (2016) Uncertainty-embedded dynamic life cycle sustainability assessment framework: an ex-ante perspective on the impacts of alternative vehicle options. Energy 112:715–728
65. Onat NC, Kucukvar M, Halog A, Cloutier S (2017) Systems thinking for life cycle sustainability assessment: a review of recent developments, applications, and future perspectives. Sustainability (Switzerland) 9(5)
66. OpenLCA (2017) OpenLCA 1.7.0.beta, GreenDaelta, Berlin
67. Pagotto M, Halog A (2016) Towards a circular economy in Australian agri-food industry: an application of input-output oriented approaches for analyzing resource efficiency and competitiveness potential. J Ind Ecol 20(5):1176–1186
68. Palmieri N, Suardi A, Alfano V, Pari L (2020) Circular economy model: insights from a case study in South Italy. Sustainability 12(8)
69. Peano C, Migliorini P, Sottile F (2014) A methodology for the sustainability assessment of agri-food systems: an application to the slow food presidia project. Ecol Soc 19(4)

70. Pelletier N (2015) Life cycle thinking, measurement and management for food system sustainability. Environ Sci Technol 49:7515–7519
71. Pelletier, N, Maas, R, Goralczyk, M, Wolf, MA (2014) 'Conceptual basis for the European sustainability footprint: towards a new policy assessment framework. Environ Dev 9, 12–23.
72. Rockstrom J (2009) A safe operating space for humanity. Nature 461:472–475
73. Sala S, Farioli F, Zamagni A (2013) Life cycle sustainability assessment in the context of sustainability science progress (part 2). Int J Life Cycle Assess 18:1686–1697
74. Sanders J (2007) Economic impact of agricultural liberalisation policies on organic farming in Switzerland, PhD thesis, Aberystwyth University
75. Sattler C, Schuler J, Zander P (2006) Determination of trade-off-functions to analyse the provision of agricultural noncommodities. Int J Agric Resour, GovAnce Ecol 5(2–3), 309–25
76. Schader C, Grenz J, Meier MS, Stolze M (2014) Scope and precision of sustainability assessment approaches to food systems. Ecol Soc 19(3):42
77. Schmid E, Sinabell F (2006) Modelling organic farming at sector level—an application to the reformed CAP in Austria. In: paper presented to international association of agricultural economists conference, Gold Coast, 12–18 August
78. Sen B, Kucukvar M, Onat NC, Tatari O (2020) Life cycle sustainability assessment of autonomous heavy-duty trucks. J Ind Ecol 24(1):149–164
79. Sua B, Heshmatia A, Gengb Y, Yu X (2013) A review of the circular economy in China: moving from rhetoric to implementation. J Clean Prod 42:215–227
80. Tarne P, Traverso M, Finkbeiner M (2017) Review of life cycle sustainability assessment and potential for its adoption at an automotive company. Sustainability 9(4)
81. UNEP (2009) Guidelines for social life cycle assessment of products. United Nations Environment, Kenya
82. UNEP (2011) Towards life cycle sustainability assessment. United Nations Environment Programme, Brussels
83. UNEP (2016) Integrated environmental assessment training manual: a training manual on integrated environmental assessment and reporting. United Nations Environment Programme, Nairobi
84. WCED (1987) Report of the world commission on environment and development: our common future. World Commission on Environment and Development Oxford
85. Wiedemann S, Davis R, McGahan E, Murphy C, Redding M (2017) Resource use and greenhouse gas emissions from grain-finishing beef cattle in seven Australian feedlots: a life cycle assessment. Anim Prod Sci 57(6):1149–1162
86. Williams AG, Audsley, E, Sandars DL (2006) Determining the environmental burdens and resource use in the production of agricultural and horticultural commodities. Cranfield University and Defra, Bedford
87. Yu M, Halog A (2015) Solar photovoltaic development in Australia—a life cycle sustainability assessment study. Sustainability 7:1213–1247
88. Zamagni A (2012) Life cycle sustainability assessment. Int J Life Cycle Assess 17(4):373–376
89. Zamagni A, Pesonen H-L, Swarr T (2013) From LCA to life cycle sustainability assessment: concept, practice and future directions. Int J Life Cycle Assess 18:1637–1641
90. Van Zanten HHE, Van Ittersum MK, De Boer IJM (2019) The role of farm animals in a circular food system. Glob Food Sec 21:18–22
91. Zurek M, Hebinck A, Leip A, Vervoort J, Kuiper M, Garrone M, Havlík P, Heckelei T, Hornborg S, Ingram J, Kuijsten A, Shutes L, Geleijnse JM, Terluin I, Veer P, Wijnands J, Zimmermann A, Achterbosch T (2018) Assessing sustainable food and nutrition security of the EU food system—an integrated approach. Sustainability 10:4271
92. Zwiers J, Jaeger-Erben M, Hofmann F (2020) Circular literacy. A knowledge-based approach to the circular economy. Cult Organ 26(2):121–141

Life Cycle Sustainability Assessment: Methodology and Framework

Shilpi Shrivastava and Seema Unnikrishnan

Abstract The main objective of this chapter is to elucidate the concept of life cycle sustainability assessment (LCSA) from the viewpoint of a life cycle. The idea of sustainability is in effect at a policy level, but it needs to be expanded in the business sector. The chapter starts with a brief introduction of sustainability followed by various approaches to perform the sustainability assessment. It also discusses how to perform the life cycle sustainability assessment using a combination of three life cycle approaches, which is a commonly used approach. The three life cycle techniques (lie cycle assessment (LCA), life cycle costing (LCC), and life cycle assessment (S-LCA)) are explained in further sections. Out of these three LCA techniques, only LCA guidelines are defined by ISO 14,040, whereas for LCC and S-LCA, the framework is still under development. Therefore, LCSA requires further improvement in the economic and social perspective, adding more accurate databases (especially for the Indian context).

Keywords Sustainability · LCA · LCC · S-LCA · Sustainability assessment · LCSA

1 Introduction

The idea of sustainability is in effect at a policy level, but it needs to be expanded in the business sector. In addition, the firms must recognize and monitor the major implications of their various processes on the environment and various stakeholders from a sustainable development perspective to execute the guidelines of the Global Reporting Initiative (GRI) [9]. Progress toward sustainability further involves modernizing life cycle assessment processes and striving for sustainable goods [14].

S. Shrivastava (✉) · S. Unnikrishnan
Center for Environmental Studies, National Institute of Industrial Engineering, Mumbai, India
e-mail: shilpi.shrivastava.2016@nitie.ac.in

S. Unnikrishnan
e-mail: seemaunnikrishnan@nitie.ac.in

S. S. Muthu (ed.), *Life Cycle Sustainability Assessment (LCSA)*,
Environmental Footprints and Eco-design of Products and Processes,
https://doi.org/10.1007/978-981-16-4562-4_3

Environmental conservation is the key to achieving this sustainable development goal. In addition to protecting the environment, evaluating economic aspects and impacts on society is also included in the outlook. The life cycle sustainability assessment system (LCSA) was, hence, proposed as it combines environmental security, economic outlook, and social equity. According to Guinée [5] and Guinée and Heijungs [6], it is possible to expand LCSA by incorporating environmental, social, and economic aspects and further expanding the boundary of a system from a micro-level (process-based) to a macro-level (economy-wide) study. These three dimensions are known as sustainability pillars, which involve the consideration of environmental, human, and economic resources, or the earth, people, and income in colloquial terms [3, 4, 7]. The dynamic relationships between the LCSA parameters also need to be understood and the mechanisms of causality between system parameters, such as economic, social, and environmental metrics, need to be studied to deepen the LCSA structure [13].

The LCSA tool is considered to be the best and offers the highest standard of evaluation among the existing environmental and sustainability tools [6]. The shift in perspective from environmental protection to economic and social protection is one of the key drivers for the introduction of the life cycle sustainability assessment (LCSA) [6] (Fig. 1).

When the 'Brundtland study' [11] introduced the idea of sustainable development to the international community in 1987, it developed a new framework for economic growth, social inclusion, and environmental conservation. The 'three pillars' interpretation of sustainability, i.e. the environmental, economic, and social aspects, is the normative model, which is well embraced by industry and often referred to as the 'triple bottom line' [10]. The conceptual formula for the LCSA framework was given by Kloepffer [8] (Fig. 2), which was further improved as presented below:

$$\text{LCSA} = \text{LCA} + \text{LCC} + \text{SLCA}$$

where

LCSA = life cycle sustainability assessment

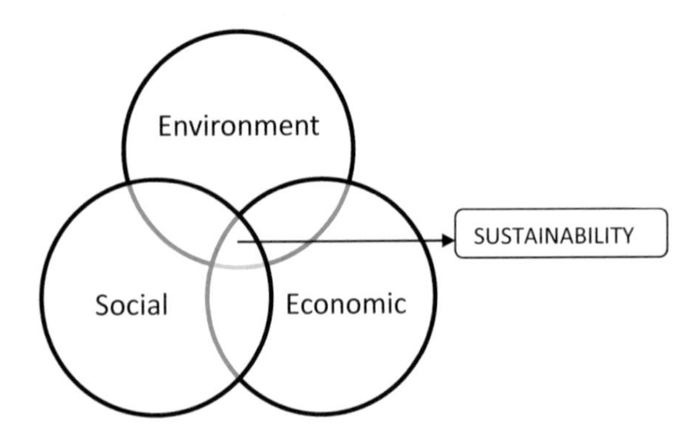

Fig. 1 Three dimensions of sustainability

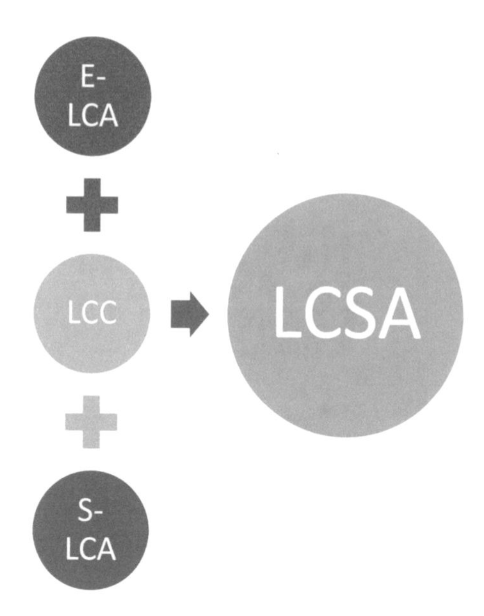

Fig. 2 Conceptual formula for LCSA

LCA = environmental life cycle assessment

LCC = life cycle costing

SLCA = social life cycle assessment.

2 LCSA Methodology

There are many approaches to perform the sustainability assessment, but the widely and commonly used methodology for evaluating the life cycle sustainability assessment is formed by combining the three life cycle techniques. Out of these three life cycle techniques, only LCA guidelines are defined by ISO 14,040, whereas for LCC and S-LCA, the methodological framework is still under development. Therefore, LCSA requires further improvement in the economic and social perspective, adding more accurate databases (especially for India) and understanding of establishing a relationship between the three dimensions of sustainability [1].

The LCSA framework consists of four phases that are (Fig. 3):

i. LCSA goal and scope
ii. LCSA inventory analysis
iii. LCSA impact assessment
iv. Interpretation

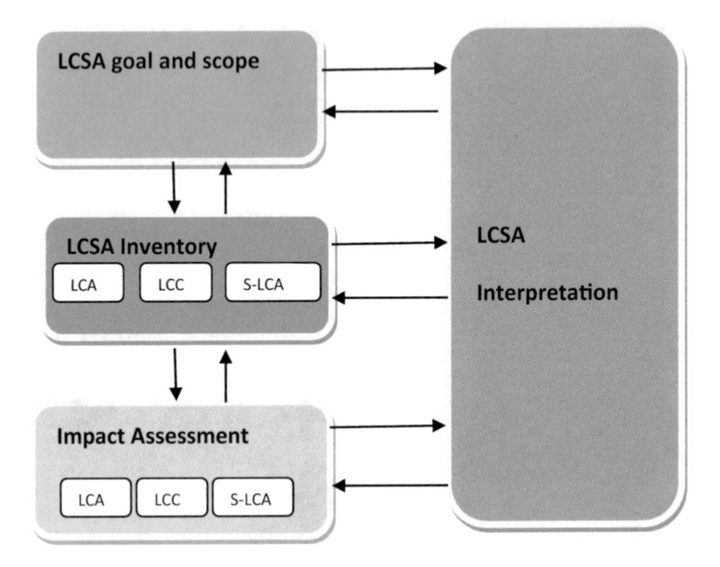

Fig. 3 LCSA Framework

2.1 *LCSA Goal and Scope Definition*

It is an important step in the LCSA process that helps to establish the study's context. It should include the following information—purpose of the study, target audience, defining the boundary, defining the functional unit, assumptions, and limitations (if any). LCA, LCC, and S-LCA (environmental) have different goals but while performing a combined LCSA, common goals and scope are strongly recommended.

2.1.1 Functional Unit

The functional unit serves as a foundation for inventory data collection and impact analysis. All the required data must be collected according to the functional unit. S-LCA does not need a functional unit because qualitative information is gathered and then translated to quantitative data for evaluation.

2.1.2 System Boundary

System boundary can be understood as a boundary or line that separates the process on which we want to focus. An attempt must be made to consider the phase, which has a maximum impact on the environment, economy, or society. When carrying out an individual assessment, each life cycle technique can have specific framework system boundaries based on its value for sustainability aspects. But there must be stages common for sustainable assessment.

2.1.3 Impact Categories

For an LCSA report, it is required that all impact categories that apply during a product's life cycle be chosen. When determining the impact categories, these should take into account the perspectives offered by each of the three approaches as well as stakeholder perspectives.

The major impact categories for each dimension are discussed below (Tables 1, 2, and 3):

2.2 LCSA Inventory

It is the most time-consuming stage of LCSA. It involves the compilation and quantification of inputs and outputs for a product throughout its life cycle (ISO 14,044:2006(E)). The inventory data must be collected as per the functional unit chosen during the goal and scope process. Tables can be generated from the collected data, and interpretations can be made. The inventory's result offers information on all inputs and outputs in the form of a basic flow to and from the environment. (Fig. 4). For LCSA inventory, both qualitative and quantitative data are taken into account.

Inventory is an interaction between the unit process and the external environment that can affect the sustainability aspects (environmental, economic, and social). Therefore, inventory data were collected individually for each life cycle attribute. It is also recommended that the inventory data must be collected according to the unit process and at the organizational level [13]. Quantitative data are collected for the LCA and LCC and qualitative data are collected for S-LCA, which is later converted to quantitative form to perform the analysis.

2.3 LCSA Impact Assessment

This step is intended for assessing the associated inventory with the environmental, economic, and social problems. This is comprised of three essential steps:

- Choosing impact categories, indicators of that category;
- The collected inventory data are grouped into particular impact categories during the classification process.
- Impact measurement, in which categorized LCI flows are described in specific equivalence units using one of several possible LCIA methodologies, and then summed to produce a complete category of effects (ISO 14,044, 2006).

Table 1 Description of impact categories for E-LCA . *Source* Acero et al. (2015)

Impact category	Definition	Impact indicator	Damage category (end point)	Unit
Acidification	Reduction of the pH due to the acidifying effects of anthropogenic emissions	Increase in the acidity in water and soil systems	Damage to the quality of ecosystems and decrease in biodiversity	kg SO_2 equivalent
Climate change	Alteration of global temperature caused by greenhouse gases	Disturbances in global temperature and climatic phenomenon	Crops, forests, coral reefs, etc. (biodiversity decrease in general) Temperature disturbances Climatic phenomenon abnormality (e.g. more powerful cyclones, torrential storms, etc.)	kg CO_2 equivalent
Depletion of abiotic resources	Decrease of the availability of non-biological resources (non-and renewable) as a result of their unsustainable use	Decrease of resources	Damage to natural resources and possible ecosystem collapse	Depending on the model: – kg antimony equivalent – kg of minerals – MJ of fossil fuels m^3 water consumption
Ecotoxicity	Toxic effects of chemicals on an ecosystem	Biodiversity loss and/or extinction of species	Damage to the ecosystem quality and species extinction	Depending on the model: – kg 1,4-DB equivalent – PDF (potentially disappeared fraction of species) – PAF (potentially affected fraction of species)

(continued)

Table 1 (continued)

Impact category	Definition	Impact indicator	Damage category (end point)	Unit
Eutrophication	Accumulation of nutrients in aquatic systems	Increase in nitrogen and phosphorus concentrations Formation of biomass (e.g. algae)	Damage to the ecosystem quality	Depending on the model: – kg PO43 equivalent – kg N equivalent
Human toxicity	Toxic effects of chemicals on humans	Cancer, respiratory diseases, other non-carcinogenic effects and effects to ionizing radiation	Human health	Depending on the model: – kg 1,4-DB equivalent – DALY (disability-adjusted life year)[2]
Land use	Impact on the land due to agriculture, anthropogenic settlement, and resource extractions	Species loss, soil loss, amount of organic dry matter content, etc.	Natural resource (non- and renewable) depletion	Depending on the model: – PDF/m^2 – m^2a
Ozone layer depletion	Diminution of the stratospheric ozone layer due to anthropogenic emissions of ozone-depleting substances	Increase of ultraviolet UV-B radiation and number of cases of skin illnesses	Human health and ecosystem quality	– kg CFC-11 equivalent
Particulate matter	Suspended extremely small particles originated from anthropogenic processes such as combustion, resource extraction, etc.	Increase in different sized particles suspended on air (PM10, PM2.5, PM0.1)	Human health	– kg particulate matt
Photochemical oxidation	Type of smog created from the effect of sunlight, heat and NMVOC and NOx	Increase in the summer smog	Human health and ecosystem quality	Depending on the model: – kg ethylene equivalent – kg NMVOC – kg formed ozone

Table 2 Description of impact categories for LCC. *Source* Simões et al [12]

Type of cost	Cost category
Initial capital costs	Equipment purchase cost
	Land cost
	Installation cost
	Fabrication cost
	Transportation cost of man and materials
	Training cost
	Alternative funding/finance auditor cost/consultant cost
	Man hour cost
	Contingency charges
	Others (if any)
Operation and maintenance costs	Comprehensive AMC
	Non-comprehensive AMC
	Supervision charges
	Insurance cost
	Others (if any)
	Comprehensive AMC
Disposal costs	Cost of demolition
	Scraping or selling assets
	Charge upon resale

The impact assessment for E-LCA can be performed by using various LCA softwares such as SimaPro, GaBi, Open LCA, etc. which can access through a complete chain for a product or a process, It takes input (raw material requirement, the electricity required, power usage, water requirement, etc.) and output (products, co-products, emissions in the air, water, land, waste generated, hazardous waste generated, etc.), and assessment result is given based on this.

For LCC, impact assessment can be done in SimaPro software, and also there are various integrated LCA/LCC frameworks available that can help in performing the assessment. LCC is generally combined with LCA because both follow a quantitative approach and the same steps. For S-LCA, there is a database available known as SHDB (Social Hotspot Database), and there are manual approaches such as Sub Category Assessment (SAM) approach, which can be used to study the social impact.

Table 3 Description of impact categories for S-LCA. *Source* UNEP report [13]

Stakeholder categories	Subcategories
Worker	Child labor
	Fair salary
	Working hours
	Health and safety
	Forced labor
	Equal opportunities/discrimination
	Social benefits
Consumer	Health and safety
	Feedback mechanism
	Consumer privacy
	Transparency
	End of life responsibility
Society	Public commitments to sustainability issues
	Contribution to economic development
	Technology development
	Corruption
Local community	Access to material resource
	Access to immaterial resource
	Delocalization and migration
	Cultural heritage
	Safe and healthy living condition
	Secure living condition
Value chain actors	Fair competition
	Promoting social responsibility
	Supplier relationship

2.4 Interpretation

The inventory analysis and impact evaluation phases are combined in the interpretation process. It is a structured methodology for defining, quantifying, reviewing, and analyzing data from the life-cycle inventory phase and/or the results of life-cycle impact evaluation. The analysis process produces a collection of conclusions and recommendations for the study. According to ISO 14,040:2006 should include the following:

- Recognition of significant issues based on the results of the LCI and LCIA phases of an LCSA;

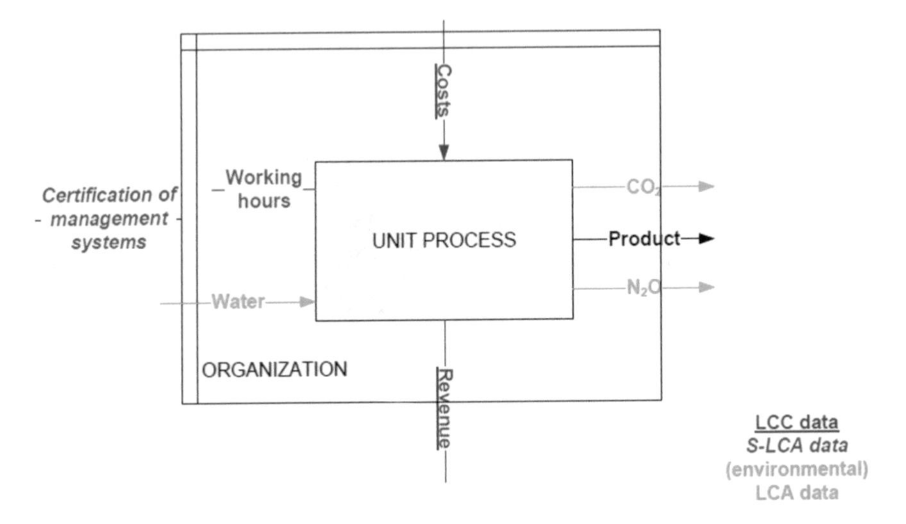

Fig. 4 LCSA inventory

- Assessment of the study using completeness, sensitivity, and accuracy checks; and
- Guidelines, drawbacks, and assumptions.

3 Development in LCSA Tool

There are few areas identified by various authors, which need more developments for advancements LCSA framework.

i. The relationship between the three pillars of sustainability is still lacking [5, 14]. One explanation for this is the existence of the metrics used in these methods, which vary from quantitative to qualitative, making aggregation difficult.
ii. It is encouraged to build up more streamlined methods that analyze the system as a whole (rather than focusing on one aspect in more detail). Software and database companies are being asked to make user-friendly and low-cost techniques available to promote more LCSAs.
iii. Adding the inventory databases from the developing countries and emerging economies for all the three life cycle techniques (LCA, LCC, and S-LCA) to make it more accurate and robust.
iv. LCSA is a new area that needs to be discussed further, and it necessitates the active participation of stakeholders and policymakers in the interpretation phase.
v. To prevent the unethical use of the tools, focusing more research on evaluating product effectiveness and sustainability.

vi. Considering the perspective of future generations in future research, adoption of LCSA strategy to prevent generational trade-offs and to take into account the Brundtland principle of sustainable development.

4 Establishing Relationship Between Three Pillars of Sustainability

To establish a relationship between the three sustainability pillars, the concept of system thinking can be used. System thinking helps us to analyze interrelationships (context and connections), viewpoints (each participant has his/her unique view of the situation), and boundaries (agreement on scope, scale, and what might be an improvement) [9]. Hence by integrating system thinking into LCSA can help in integrating and establishing the relationship between three sustainability aspects. It will also act as a decision-making tool and will help in effective policymaking. It consists of the following steps:

4.1 Identifying Variables of the Process Chain

The process chain must be studied from a system dynamics perspective, taking into account the complex and causal relations between the environment, economy, and society. For each sustainability aspect of the process chain, the variables must be described based on the LCSA study performed. The variables can be identified from the impact assessment results, which show the most affected impact categories.

4.2 Creating a Causal Loop Diagram

A causal loop diagram lets one understand how different variables in a system are interrelated. The diagram is composed of a series of nodes and edges. The nodes represent the variables and the edges are the links that represent a relationship between the two variables identified for the study. The causal loop diagram presents the most important relations between identified parameters of the system and explained how system thinking can be used to present a clearer view of the interest system's underlying mechanisms and their impacts on different aspects of sustainability.

The causal loop diagram is composed of four essential components.

i. Variables: The initial step in developing a causal loop diagram is identifying variables. For this study, the variables were identified from our LCSA study of crude oil in India.
ii. Drawing Links: The next step is to draw links between the identified variables, filling it with the verb, and determining how one variable impacts the other. The

links are labeled as '+' '–' or 's' 'o'. If variable B goes in the same direction as variable A, it will mark the relation between variable A and variable B with "s" (or "+"). If variable B moves in the opposite direction of A (i.e., as A increases, B decreases), the relation between A and B should be labeled with an "o" (or "–").

iii. Labeling the loop: In this step, we identify the behavior of the loop. There are two fundamental types of causal loops in systems thinking: reinforcement and balancing. If a change in one direction is amplified by any further changes, then it is a reinforcing loop. In balancing loops, alternating loops counter-shift in one direction with a shift in the opposite direction.

iv. Talking with loops: By connecting various loops, we create a concise story for a particular problem. To ensure this, we must follow the links and capture the loop behavior.

5 Conclusion

Startups, government organizations, international cooperation agencies, and other societal entities will all benefit from LCSA in their efforts to generate and use more sustainable goods. This entails cost-effectively minimizing environmental pollution and the conservation of natural resources while also contributing to social well-being.

The discussed LCSA approach can be used to determine the sustainability of all goods and processes, providing useful data to policymakers. Since the LCSA framework is still in the developing phase, more developments are needed.

The integration of system thinking with sustainability assessment will help in understanding the system as a whole, the interaction between various subsystems, and identifying uncertainties and dynamic complexities.

Strong and dependable science-based techniques are needed to produce expertise in the field of resource efficiency and then translate a deeper understanding of the commodity system into action to achieve the aim of a green economy with sustainable consumption and production patterns.

References

1. Costa, D., Quinteiro, P., Dias, A. C. (2019). A systematic review of life cycle sustainabilityassessment: current state, methodological challenges, and implementation issues. Sci Total Environ 686:774–787
2. Cristina, A. R., Andreas, C. (2015). Impact assessment methods in life cycle assessment and their impact categories. *1.5.4*.
3. Elkington, J. (1997). Cannibals with forks: The triple bottom line of 21st century business, Capstone: Oxford.
4. Foolmaun, R. K., Ramjeeawon, T. (2013). Comparative life cycle assessment and social life cycle assessment of used polyethylene terephthalate (PET) bottles in Mauritius. Int J Life Cycle Assess 18(1):155–171.

5. Guinée, J. (2016). Life cycle sustainability assessment: what is it and what are its challenges?. In Taking stock of industrial ecology (pp. 45–68). Springer, Cham.
6. Guinée, J. B., & Heijungs, R. (2011). Life cycle sustainability analysis: Framing questions to approaches. *Journal of Industrial Ecology, 15*(5), 656–658.
7. Kajikawa Y (2008) Research core and framework of sustainability science. Sustain Sci 3(2):215–239
8. Kloepffer W (2007) State-of-the-art in life cycle sustainability assessment (LCSA) life cycle sustainability assessment of products. Int J LCA 13(2):89–95. https://doi.org/10.1065/lca2008.02.376
9. Meadows, D. (2006). Green building materials – a guide to product selection and specifi cation. 2nd edition. Wiley, Virginia, USA.
10. Onat NC, Kucukvar M, Tatari O, Egilmez G (2016) Integration of system dynamics approach toward deepening and broadening the life cycle sustainability assessment framework: a case for electric vehicles. Int J Life Cycle Assess 21(7):1009–1034
11. Shrivastava, S., Unnikrishnan, S. (2021). Life cycle sustainability assessment of crude oil in India. *journal of Cleaner Production, 283*, 124654.
12. Simões, C. L., Pinto, L. M. C., & Bernardo, C. A. (2012). Modelling the economic and environmental performance of engineering products: a materials selection case study. *The International Journal of Life Cycle Assessment, 17*(6), 678–688.
13. UNEP, SETAC. (2009). Guidelines for social life cycle assessment of products. United Nations.
14. Zamagni, A. (2012). Life cycle sustainability assessment.

Application of Life Cycle Sustainability Assessment to Evaluate the Future Energy Crops for Sustainable Energy and Bioproducts

R. Anitha, R. Subashini, and P. Senthil Kumar

Abstract Currently, Life Cycle Sustainability Assessment (LCSA) methodology is widely used to determine possible environmental impacts in the sustainable fuels and energy sector. It is a deep-rooted tool to afford data-driven investigation of environmental impact assessment. Among the energy crops, specifically, the perennial grasses and trees might contribute significantly to the mitigation of global environmental problems in energy safety and climate change, provided if high yields can be attained. The nonfood low-cost perennial energy crops like Salix, Miscanthus, switchgrass, and giant cane grass are considered commercially important due to their high-yielding capacity, can grow in marginal land type, minimum requirement of input needs, and more ground cover. These crops would address several environmental issues such as involvement in the reduction of greenhouse gases and energy either for bioenergy or biomaterials and encouraging social benefits specifically in rural areas. But its economical utilization is compromised since their cost of production is influenced by yields. The present chapter discusses the OPTIMA project which is dedicated to the farming of perennial crops such as giant reed (*A. donax* L.), miscanthus (Miscanthus × Giganteus), and switchgrass (*Panicum virgatum* L.) in minimal nutritional soils in the Mediterranean region. This chapter also briefly describes the sustainability assessment of the perennial energy crop cultivation with reference to economic, environmental, and socioeconomic benefits and elucidates the validation on cultivation and utilization of perennial grasses to predict the advantages toward sustainability.

R. Anitha
Assistant Professor, Department of Biotechnology, Hindustan Institute of Technology and Science, Chennai 603103, India

R. Subashini · P. S. Kumar (✉)
Department of Biomedical Engineering, Sri Sivasubramaniya Nadar College of Engineering, Kalavakkam 603110, India
e-mail: senthilkumarp@ssn.edu.in

P. S. Kumar
Department of Chemical Engineering, Sri Sivasubramaniya Nadar College of Engineering, Kalavakkam 603110, India

S. S. Muthu (ed.), *Life Cycle Sustainability Assessment (LCSA)*,
Environmental Footprints and Eco-design of Products and Processes,
https://doi.org/10.1007/978-981-16-4562-4_4

Keywords Perennial grass · Future energy crops · Sustainability · LCSA · Miscanthus · Switchgrass · Bioenergy · Environmental impact

1 Introduction

The energy policy of India is basically demarcated by the country's increasing energy deficit that results in the increasing emphasis on the development of alternative energy sources predominantly toward renewable energy system. In recent years, India's achievement in the energy sector is outstanding and also the Government of India is executing improvements headed for a safe, reasonable, and justifiable energy structure to rule a healthy monetary development. In the United Nations' Climate Summit held during September 2019, it has been proclaimed that to achieve the renewable energy towards a target of 450 GW by 2030 [1]. To project and improve such a structure, numerous works were directed, nevertheless all these studies highpoint the necessity of a possible biomass production of about 35–50% of total energy depletion [2]. There are many motives to illuminate why the biomass is very striking for energy systems exclusively without fossil energy [3] and the main advantage of which is capable of storage, which is the main reason to maintain the changing energy buildup from recurrent bases such as wind and solar.

The availability of adequate primary and ancillary energy sources to tackle the requirements of the people is becoming a subject of matter and currently, sustainability is receiving more attention globally. Among the available energy sources, maximum energy is supplied from the fossil source but they are limited, nonrenewable yet the practice of which increases severe environmental issues. In reality, the increased emission of greenhouse gases and additional contaminants are due to the exhaustion of the identified reservoirs of petroleum, natural gas, and coal at a fast rate. The manufacturing of a range of multifaceted supplies like plastics, cleansing agents, solvents, resins, lubricants, epoxy resin, fibers, and elastomers are produced from the source supplies like alkenes and aromatics, these base materials are produced from fossil resources. Hence, the search for alternative energy from various means for improving the stock assembly become a key phase to stop energy deficiency, weather variation, and exhaustion of fossil fuel materials.

Yet biofuel signifies a natural or renewable energy source but it is not unlimited in supply, since it is naturally replenished, it has been measured as an alternative raw material to deliver renewable energy in near future. The ancillary energy sources are generated from these raw materials such as agricultural and forestry residues, waste materials generated from industries and municipal solid waste (MSW). Unlike first-generation biomass which has been obtained from the wholesome food crops like wheat, corn, sugarcane, barley, sunflower, potato, and soybean, the second-generation biofuels are produced from plant dry matter (Lignocellulosic waste materials) like switchgrass, cassava, jatropha, straw, and wood. Using the biomass and residual waste as a principal source of biofuel production is a favorable scheme for eco-friendly waste disposal. Light energy is fixed and stored in plants through the process of photosynthesis and in see grass, by means of sugars, besides subsequently the biomass created are transformed into different kinds of biofuels such as solid, liquid or gas-mixture and heat. Additionally, the resulting bio produces obtained from it are well-matched with the prevailing systems and are recyclable. Perennial grasses, like *Arundo donax* (L.) and *Miscanthus* spp. are considered as cost-effective and amicable crops, it is possible to convert the abundant biomass produced into manufacturing of power (dense and secondary biofuels), pulpwood, and biocompatible materials [4]. The higher productivity of these crops is due to their low-input costs and requirements, low water needs, and constructive ecological effects (e.g., possible by means of carbon sink, reclamation and riddle structures, by increased water and nitrogen use efficacies). Because of these reasons, the present study focus on sustainable energy production using only the perennial energy crops. However, the growing pressure on renewable resources increases the competitiveness for farmland, emphasizing the energy against nutrition quandary and the modification in the land practice discussion [5].

This chapter aims to give a recap of the current status of investigation and application concerning life cycle sustainability analyses pertaining to the evaluation of perennial energy crops and also to consequently identify research issues. In addition, to encounter the renewables or biofuel purchase, ecological conservation has to be examined. The present paper delivers a broad assessment of researches on the self-sufficiency of long-term grasses cultivation and application with main focus on ecological, monetary, and social-class effects. Additionally, through a combined approach, possibilities for inhibiting disadvantages and intensifying advantages are specified to offer innovative understandings into the upcoming progress of the energy crops in an ecological farming milieu.

1.1 Perspectives of Life Cycle Sustainability Assessment

Over the last three decades, the environmental life cycle sustainability assessment (LCSA) has been recognized and from the late 1960s and early 1970s, the LCSA studies are documented and during this period, environmental issues like solid waste, toxic waste control, and energy efficiency turn out to be the matters of general public concern. In an earlier study led by the MRIGlobal Technological and Scientific Research, in 1969, enumerated the resource supplies, discharge loadings, and leftover flows of various beverage containers. Together with numerous developments, it cleared the commencement of progress on LCSA since we recognize it nowadays [5–7]. The time interval 1970–1990 encompassed the periods of commencement of LCSA with broadly deviating tactics, terms, and results. LCSA was done with various methods without a common hypothetical agenda and the results attained varied significantly, still when the content of the research were same [8]. During 1990s, an outstanding development was observed in the scientific and organization actions worldwide, which among the new things is revealed in the number of

LCSA books and manuals created. The period from 1990 to 2000 revealed concurrence and coordination of approaches by SETAC's organization and ISO's regularization events, as long as a consistent outline, terms, and platforms for discussion and management of LCSA methods have been achieved. LCSA has become progressively a part of strategy papers and legislature, primarily concentrating on wrapping in this period and in this time the scientific field of industrial ecology (IE) rose [9]. LCSA was gradually used as an instrument for backup strategies and performance-based guidelines. During this period, the carbon footprint based on life cycle principles were recognized globally. Further, during this period, LCSA methods were explained in detail, but it inappropriately gives rise to discrepancy in the methods again. After that, several life cycle valuation (LCV) and social life cycle valuation (SLCV) methods were predicted. Distinctive origin of LCSA was distributed by the environmental constituent but with these LCSA widened itself from a simply ecological LCSA to an additional complete life cycle sustainability assessment (LCSA). This widening is reliable with advances in IE and the durability with the three-pillar method are the primary drives [10].

1.2 The Energy Crops

The crops grown primarily to offer raw materials for energy industries are called energy crops. Since, biomass is an inexhaustible fuel, apart from its existing accessibility, the annual yield should be considered. The annual crops breeds each year then perishes and which is considered as the entire dead plant matter produce. Whereas the everlasting plants such as trees which do not decease each year and which is considered as the annual growth of the plant. The other significant factor that impacts price is the fuel heating value and the section of the total biomass accessible for energy creation.

1.3 Perennial Energy Crops

Bioenergy is able to be carried through a range of crops, harvesting systems, and adaptation skills and it is demanded that, challenged with the mutual competition of nutrition and energy safety, little response perennial crops are ecologically greater to annual crops. These are calledblossoming plants with a multi-year life cycle of, typically roughly, 10–15 years. The examples comprise numerous familiar vigor crops like cardoon, miscanthus, switchgrass, giant reed, and reed canary grass. The solar capture by plants reveals that the photosynthesis of C4 persistent energy crops is as effective as photovoltaic devices [11]. The key benefit of perennial crops is plowing, and implanting are not necessarily done for each year but, the yearly watering, manures, insecticides are essential. The preliminary farming cost can be separated during the period of the life of crop. But the prolonged life cycle might appear to farmers as a longstanding promise with insecure forecasts, particularly once there are no clear market frameworks for the end product [12].

The perennial grasses are the right candidate for biofuel production because of their higher yield potential, increased polysaccharide content, and also their positive social and ecological advantages [13]. It has abundant potential for ancillary biofuel production, and still the above crops have some benefits over annual crops based on the agricultural inputs, profits, production costs, food security, reduced GHG (greenhouse gas) releases, and environmental sustainability. The root systems of perennial grasses are widespread in nature and can possibly bind to the soil which results in preventing erosion and aids in straining heavy metal contaminants from wastewaters [14]. Numerous studies specify that the soil erosion influenced by water will be decreased predominantly when cardoon, miscanthus or switchgrass substitutes old yearly crops [15]. These crops have been reported to decrease the N and P loss to shallow and groundwater than the arable land uses. Besides that, the perennial grasses are having the ability to significantly upsurge the carbon-based soil quantity especially in earths with earlier exhausted C levels. The perennial crops might also be used for phytoremediation of places polluted with thrash metal and for wastewater treatment, farmland wastewater, mud, and landfill leachate. Loss of nitrogen from farmland lessens biodiversity of ordinary environment, contaminates potable water, thus disturbing human health, and pays for climate change which leads to global warming. Nevertheless, to encounter the targets of the EU Water Framework Directive, still declines are needed. Hence to achieve the goal, the crops that should positively influence the environment have been recommended. These crops are acting as carbon sinks since, it helps in reducing the atmospheric GHG significantly when compared to the annuals like food grains, there will be a fall in N_2O- emissions. The carbon storage capacity of these grasses is 1,565 tons of CO_2- corresponding per hectare. Making new sources of income is possible for the countryside communities, as well as occupation ventures, through cultivating these crops on minimal lands, deprived of negotiating current food crops. The regional economic structures are developed and it brings out the improvement in the learning, training, and supports provided for farmers [16]. Some of the agronomic features of perennials like blossom sterility, great early fitting price, comparative little produce mechanization, extraordinary moisture throughout harvest, and high ash percentage are substantial flaws. Another significant disadvantage is that the use of more water when compared to ancient crops and grassland. Also, there is a worry about protection of the exposed farming landscape, where 4–6 m tall green walls of perennial crops hindering the sight hence, possible substantial influence on landscape problem is there. Additionally, if huge regions are implanted, it might lead to monoculture, and lastly, root penetration and hydrological impact may adversely disturb the archeology. Perennial grasses need a long-term land commitment by the farmers. Some of the perennial crops are exclusively planted using rhizomes, which makes it more costly than seed farming. Therefore, they have lower farmer/public acceptance as equated to the annuals.

1.4 Resource-Utilization-Efficiency (RUE)

Everlasting vigor grasses are extraordinary RUE crops with respect to sunlight, water, and nutrients, and are low-input challenging. According to CAM (Crassulacesan Acid Metabolism) plants, a RUE of 1.6–5.0 g MJ^{-1} of diverted vigorous energy (iPAR) was estimated in switchgrass [17], 1.1–2.4 g MJ^{-1} in Miscanthus grownup, in Texas, under irrigation [18]. While giant reed followed the C3-pathway, its CO_2 absorption, light capture, and dry matter yield is comparable to that of CAM plants. In switchgrass, a typical prompt WUE of 6.5 for low-lying and 4.4 μmol CO_2 mmol H_2O^{-1} for highland ecotypes has been stated. Such ecotype variations were mostly associated with various water necessities, external structures, and productive traits [19]. The increase WUE of switchgrass was associated not only with its CAM metabolism, but also with its root length thickness and increased water uptake ability [19]. Although a high RUE has been related to reed canary grass, works on RUE and WUE are missing, since the growing environments are not restrictive for light and water. At the same time, numerous researches have explored the result of N on reed canary grass, with conflicting outcomes. There is not a proper yield responses observed on N fertilization above 100 kg N ha^{-1} [20]. Smith and Slater [21] established that no important effect with N contents up to 87.5 kg ha^{-1}, either with inorganic or organic manures. Surely, profit increases were recorded in Germany with N source up to 163 kg ha^{-1} but with a related reduction in NUE. While there is a significantly greater everlasting grass species compared to that of annuals, they are frequently fewer receptive, or even become inefficient at high N rates because of nutrients cycling. As far as the CAM plant Miscanthus is concerned, a regular and characteristic higher NUE was recorded when compared with the C3 giant reed (442 vs. 382 g g^{-1}, respectively) grown in the Mediterranean area. [22] accomplished characteristic effects of N fertilization on NUE in giant reed at the first and the second growing period, but not during the third one. NUE was exploited at a manuring rate of 60 kg N ha^{-1} $year^{-1}$ as compared with 120 kg N ha^{-1} $year^{-1}$. In switchgrass, NUE reduced as well when N was elevated from 45 to 180 kg ha^{-1}. Comparable tendency were also achieved by Lewandowski and Schmid [20] with Miscanthus and reed canary grass signifying that lower N fertilizer rates for the C4 Miscanthus than for the C3 reed canary grass to exploit NUE.

1.5 Ecological Paybacks

The noteworthy sustainability features of evergreen perennial grasses are the lignocellulosic assembly of cell walls that protect resistance toward pests and diseases. These crops have a little natural enemies and their farming is pesticide-free to date. Like other perennials and deep-rooted crops, deep soil spadework is vital to confirm long-lasting performances. Yet, several works stated extraordinary growth under least agronomic techniques [16]. Once proven, the control is fairly simple since perennial grasses struggle fairly sound with weeds at regrowth and also they need low or no-input, so, only dry matter harvesting is directed yearly. Grown up plants increase the soil structure, its constancy, and health; further, they deliver a canopy cover with an advantage for biodiversity.

2 Methodological Characteristics

2.1 *Life Cycle Sustainability Assessment*

Life cycle assessment (LCA) is an ecological organization means and accounts ecofriendly features attributes and possible conservational effects of a product. An extensive choice of impact groups is enclosed with a complete depiction of the invention's ecological references. The evaluation comprises the invention's complete life process from raw material procurement over manufacture, usage, end-of-life treatment, reusing, and end clearance. This aids to evade a flowing of ecological loads among life process phases, among geographic provinces, or among influence groups. LCA is globally consistent over ISO standards and others can support in recognizing chances to recover the ecological enactment of crops at numerous facts in their life cycle. The above ISO morals describe four stages in an LCA study (Fig. 1) and this agenda, yet left the single specialist with a variety of options, which can

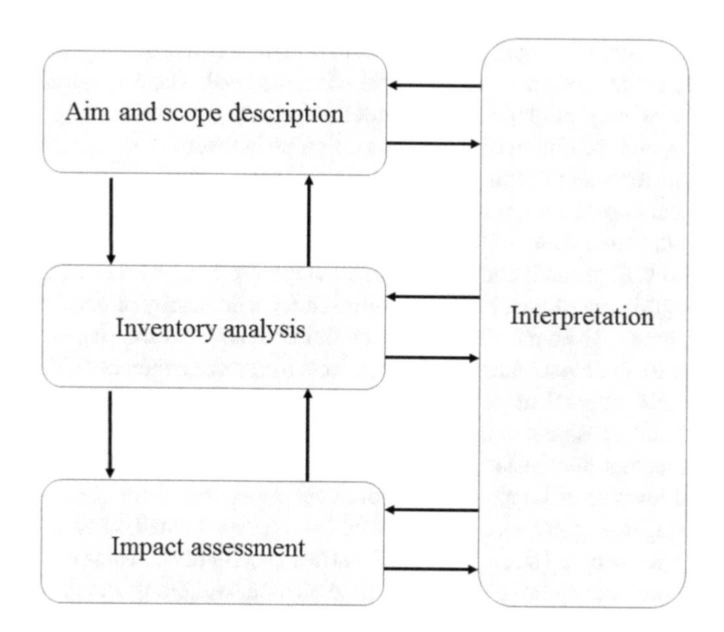

Fig. 1 Design of four segments of a life cycle valuation (according to [23, 24])

disturb the acceptability of the outcomes of an LCA study. Whereas flexibility is vital in replying to the big variability of questions lectured, additional management is required to confirm reliable and superior LCA studies. The LCA Handbook is a sequence of methodological papers that offer comprehensive direction on all stages essential to conduct an LCA. Also, it stipulates in which choice setting suppleness or firmness concerning these guidelines is more imperative.

The fourth stage of LCA influence valuation proceeds over 4 steps. In the first 2 steps, the assortment of impact groups and organizations are compulsory. For offering a complete depiction of the invention's ecological effects, typically a varied choice of impact groups is covered. The subsequent two steps such as standardization (3) and weighting (4) are yet to be elective. While well-known and appropriate for the valuation of worldwide and supra-regional ecological impacts, typical LCA practice to date is not able to speak native and site-specific influences on ecological issues such as biodiversity, soil, and water. As long as organizational progress into this track are still continuing, standard LCA should be complemented with an estimate of native and site-specific impacts based upon features hired from Environmental Impact Assessment (EIA) and Strategic Environmental Assessment (SEA).

2.2 *Ecological Impact Assessment (EIA)*

EIA is a valuation technique to determine the likely ecological outcomes of a future project. EIA scrutinizes the expected ecological results and regulates the reputation of these effects, in both the small and extended period. It emphasises indigenous ecological effects, figures together and evaluated with that level. The ecological influence examination of crop production needs good information of the farming processes, the supplies, and the output of the numerous crops in diverse weathers, varieties of soil, and approaches of farming. There is not an overall list of standards to measure the ecological impact nor a general report of approaches to be used. Setting the environmental standards is part of the EIA development and generally, standards address releases into soil, ground, and surface water and air, effects on the living environment and well-being of people in the settings, effects on nearby biomes, and effects on cultural assets. Even though EIA can be more vivid, it is essential for collective information to abbreviate many records of facts into more coherent evidence about probable environmental impact. To ease a straight assessment, parameters can be regularized and translated to the same degree. A modest form of normalization can be used, all factors are translated into a number between 0 and 100, for example, with 0 being the lowermost impact and 100 the uppermost impact for each group [25]. In the last step, the scores on the various meters can be weighted. Crucial weighting features is a value-based statement, which carries uncertainty and bias to the study at hand. Some authors approve that, every time applied, weighting should imitate the relative status of the influence categories in the structural setting of the study [26].

2.3 Financial Study of Crop Breeding

Financial study scrutinizes the effectiveness and monetary durability of schemes to evaluate the desirability of subsidy other venture chances. In precise, the monetary investigation of eternal crops necessitates the valuation of entire prices and incomes made in each and every year through the financial life of the studied crop and the essential dimensions and control of the essential venture.

Reduced money movement approaches [27] can be implemented for the study of multi-annual crops since financial examination desires to utilize the financial performance of those schemes during their financial life (life cycle financial study (LCEA) [28]. Revenue and expenditures of agronomic projects differ pointedly from time to time because of the physical progress of multi-annual farms and the varying requirements and profits that are indicated by farming practices. Scheme success is designed as the variance among revenue and expenditures. Income is received largely from the trade of goods and facilities. Expenditures comprise of groups like man power, technology and tools, source materials, hired services (out-sourcing), property payment, monetary and tariff expenditures, etc. Pay and costs are not continuous throughout the monetary life of the farms, and as a consequence, productivity diverges from year to year. Farming projects to hurt sufferers are not rare throughout the initial years of the crop and relish decent incomes later, when the farm is established and produces are high. Typically, viability standards and catalogs are stated for established farms besides are lost the gathered losses during the initial years, which are the most significant for the agriculturalist or the capitalist. While output metrics are normally the greatest extensively used and simply agreed events of performance, they don't deal with the depositor whole information, since they don't disclose vigorous cash influx and discharge particulars, which may perhaps be the most vital. The study of the scheme, cash flow, is vital, mainly for the purpose of assets planning and venture assessment, when we want to compare the current worth of net arrivals to the capitalized amount, which is typically paid upfront. Because of the time worth of money, the stream of prices and incomes of farming projects is hard to evaluate with other events with diverse cash flow designs, except money standards are spoken in some common denomination. Discounting of future financial flows is common in financial assessment, since it allows the calculation of one value figure, the current value, which proves the entire stream of cash flows. In the operational method, each crop is scrutinized for the whole of its valuable life. To approximate costs, agricultural production is fragmented down into single operations or actions and the necessities of each activity are documented and dignified with respect to human or machine hours, volumes of raw materials spent, rental, etc. The original venture is clearly recognized and estimated. Farm financial records do not typically recognize the full cost of farming production, maybe because of the absence of promise and figures on credited costs, such as family labor, own land, etc. For monetary analysis, these items should be predictable at their occasion price to classify the net revenue credited to the project. Economic practice needs the decline of the project into a number of processes or activities, which adequately define crop installment, cultivation, harvesting, and storage activities. Each operation is categorized by its timing and its requirements for land, labor, equipment, and materials. Seasonality is important if topmost labor, machinery, and water needs have to be recognized.

All cost stuffs are initially dignified in physical quantities, for instance, land area, labor and machine hours, amount of raw material, quantity of fuel required, etc. This delivers a price dimension system self-determining of prices of resources, at least in the small run. The required amounts of factors of crop production and raw materials are then increased by their consistent prices to calculate the total cost in financial terms. Mechanical equipment may be involved, if own machinery is insufficient or absent. When hired, its cost is equivalent to the rental paid and the yearly cost of own equipment is the sum of devaluation, attention, care, insurance, labor, and fuel. If separated by hours of operation per year, it provides an approximation of the hourly price of the equipment. The land is a vital factor of agricultural production and in most cases the main cost item. The cost of agricultural products may be meaningfully amplified if established on high-cost land and vice versa. Therefore, land cost must be cautiously appreciated in all agricultural projects. If there is a properly modest market for land, one may assume that its rent sufficiently reflects its real cost. But, if there is no market, the cost of land is not just identifiable. In such cases, one needs to approximate its chance cost as communicated by the net economic output of existing or usual land use. Marginal land rent is much more interesting to estimate because its opening cost is very site specific and because of likely changing subsidization. Labor is typically provided by the farmer and his family, but it may also be appointed, particularly during peak labor demand, e.g., at planting or harvesting times. Hired labor in most cases has a market-specified rate, which can be used in the analysis. Allocated labor cost should be mainly measured at its opening cost, i.e., the amount of income mandatory for shifting family labor from current activity due to the needs and supplies of the project. In general, when there is no market for a product, the opportunity cost of the pertinent issue or production should be used to evaluate the cost of inputs. Opportunity costs should imitate circuitously the market values. For example, made disposable inputs should be valued at the cost of buying the input from off-farm. Likewise, capital facilities provided by the owners of a specified enterprise should be valued at the cost of gaining these facilities from another source in a market deal. To review the findings of economic analysis, it is useful to estimate economic guides, which reveal the possibility and feasibility of agricultural savings. Usually recognized directories also provide a basis for the contrast between alternate venture plans. The basic monetary indices suitable for financial analysis of crop sustainability are the following:

1. Return on total assets: This percentage demonstrates how professionally assets are making incomes (before interest and tax) as a fraction of total assets. It exhibits the profitability attained by each INR of the assets obligatory by the project.
2. Reimbursement period: This is one of the modest and most extensively used venture assessment keys. It deals with the number of years wanted for net project advents to pay back the initial venture. In the case of multi-annual agricultural

projects, the initial investment includes the usual land, machinery and apparatus, buildings and structures, and the expense of buying and fitting the plantation. This modest type of key does not need ignoring future cash flows. It shows not only the speed of capital recovery, but also the extent of danger, since the slighter the payback period, the lower the risk.

3. Net current value (NCV): This is the existing value of the stream of net cash flows (inflows minus outflows) through the monetary life of the plantation. This is a degree of the economic attraction of projects and positive NCVs entitle projects capable of making commercial excess after having paid all project costs and expenditures, including the initial investment expense. The mathematical formula for the calculation of NCV is presented following:

$$\text{NCV} = n \sum t = 0[\text{CF}_t \times (1 + d) - t]$$

where CF_t is the net cash flow of year t (influxes minus losses), CF_0 is the net cash flow of year 0, classically the initial venture outflow (negative), CF_n is the net cash flow of year n, including likely land renovation costs or positive terminal value of the plantation, n is the number of years of the economic life of the plantation, d is the discount rate.

4. Internal rate of return (IRR): This is the reduction rate (d) for which NPV = 0. The greater the discount rate, the lesser is the NPV. Thus, the IRR designates the maximum rate of return (ceiling) that the project can attain, or the maximum interest charge of capitalized capital outside which the project is not financially fulfilling.

2.4 Demographic/Socioeconomic Analysis

The socioeconomic features are interweaved with the economic analysis and quantify the socioeconomic influence concentrating on both quantitative (occupations, direct, indirect, and induced) and qualitative (influence to rural economy, local implanting, and nearness to markets) limits. Generally, methodological tactics pool qualitative and quantitative valuation and assess individual impacts in two categories, i.e., service effects and communal sustainability (Table 1).

After considering the recurrent crop making and use, the jobs intended are net created jobs (formed jobs minus lost jobs due to substituted earlier uses of the land). The principle can be additionally specified by the following pointers: Straight jobs: The subsequent value chain steps need service that could be comprised of the dimension of direct jobs formed:

1. Influence of rural economy: Service is the main subject in rural economies. Certain value chains may bring more local job creation, inspiring the rural economy, however, other value chains might be more engaged to large-scale industry, often in the hands of international players/multinationals.

Table 1 Analysis of socioeconomic impacts

Parameter	Qualitative factors	Quantitative factors
Class	Occupation effects	Socio-sustainability concerns
Standard	Job (standards or preservation)	Influence on the country's economy Native implanting and nearness to marketplaces
Indicator	Shortest job counterparts for the price chain Secondary job counterparts for the price chain Net added prompted jobs	Qualitative (high, modest, and low) Qualitative (high, modest, and low)

2. Native implanting: The ability of the native economy to progress and function a full value chain or part of it (in the OPTIMA case the manufacture of perennial crops).
3. Vicinity to marketplaces: The pointer states the variation between a more local method with low distances on the one side, and on the other side a more international/industrial approach, where the feedstock is transported to large manufacturing places or to harbor areas to be moved. The first four principles are typically linked with biomass made outside Europe and imports. They frequently worry about life and functioning conditions in poor countries with low safety standards, and even if companies are not straightly involved, their source chains maybe unrecognized by themselves possibly will make them accountable for obvious mistakes. High-risk possibilities for forced labor in Europe are newsworthy according to a greatest hot report custom-made by the UK-based Joseph Rowntree Foundation. Managements should struggle to start a consistency of the raw material policy with human rights obligations, risk valuation for human rights violations for trade agreements of the European Union with third countries, building support programs for projects in foreign countries dependent on due perseverance for human rights, beginning raw materials partnerships with foreign countries with the valuation of significances for human rights, and supportive governments in foreign countries to pass issues such as the right of codetermination, and in specific, the right for free, early, and informed contract of native people to projects about their own environment and living. The same source approves to enterprises, among others, that they participate human rights moralities in their own policies at the highest organization level, demanding for human rights morals in supply contracts, launching self-governing assessing with a focus on human rights risk assessment, evolving certification that addresses all relevant human rights standards, launching a material data bank including all pertinent information for use in suppliers valuation and provisions for tender formulations, and creating a broadcasting system on one's own rehearsal and efforts to gain guidance for the supply chain with regard to human rights.

2.5 Ecological Facets of Perennial Crops Production and Use

The most vital people's top ten difficulties for the next 50 years are energy, water, food, environment, poverty, terrorism and war, disease, education, democracy, and population. The global population will continue to grow, yet it is likely to plateau at some 9 billion people by roughly the middle of this century [29]. And as the population grows, so too does the demand for land and energy which, together with climate change, will further hinder agriculture's capability to produce enough food to withstand society. Hence, several investigators approve that agriculture is the major risk to biodiversity and ecosystem functions of any single human activity. To figure from the 2005 synthesis statement of the United Nations' Millennium Ecology Valuation Program Farming frequently has a negative impact on the delivery of services. For example, cultivated systems incline to use more water, upsurge water pollution and soil erosion, stock less carbon, produce more greenhouse gases, and support meaningfully less habitat and biodiversity than the ecologies they replace. Presently, more than two-thirds of worldwide cropland is sown to monocultures of annual crops, much of the land meant for annual crops is previously in use; and manufacture of nonfood goods progressively contests with food making for land. The best lands have soils at low or reasonable danger of degradation under yearly grain manufacture but makeup only 12.6% of the global land area. Associating more than 50% of the world's population is another 43.7 million km of marginal lands (33.5% of global land area), at high risk of deprivation under annual grain production. With more land worldwide having been rehabilitated from perennial to annual cover since 1950, than in the earlier 150 years, the area involved by annual species remains to expand, the risk of soil degradation looms larger. This recent expansion of cropland has made it more and more necessary to apply chemical fertilizers and pesticides, which disrupt natural nutrient cycles and erode biodiversity. Perennial crops would discourse many agricultural problems as well as considerable ecological and economic benefits, relative to annual crop species, they can harvest more ground cover, and achieve longer growing seasons and more extensive root systems, which make them more viable against weeds and more operative at arresting nutrients and water. Therefore, it can be used in reducing soil erosion, lessen nutrient leaching, sequester more C in soils, and deliver continuous locale for wildlife. In addition, combinations of species in intercrops or polycultures have the possibility to improve the performance of a cropping system in terms of yield, nutrient cycling competence, and pests control. In a field trial surrounding 100 years of data collection, annual crops were 50 times more vulnerable to soil erosion than perennial grassland crops, and annual grain crops can lose 5 times as much water and 35 times as much nitrate as perennial crops. Though biomass-based organizations require un-replenishable energy for the cultivation, transport, and adaptation to bioenergy, the energy balance associated with the whole life cycle, restrained by the ratio of nonrenewable energy input/energy output, is usually lower than 1, meaning that, it uses less un-replenishable energy than the energy it provides. An earlier study reported that those lignocellulosic crops give superior energy savings per hectare than wood chips and wood pellets or kenaf, hemp, and cardoon, by means of the same energy-generating technologies, and parallel to those stated for sweet and fiber sorghum [30]. Biomass use as energy or materials is restrained a carbon saver over its life cycle since, carbon has been arrested from the atmosphere and has been photosynthetically converted into bio-matter using solar radiation, water, and external inputs. Though, a portion of CO_2 is released during the cycle of biomass production and use: external fossil fuel inputs are vital to grow and harvest the feedstocks, in transport and in dispensation and handling of the biomass. Other gases such as N2O and CH4 can also pay to the greenhouse effect (ascribed to the nitrification and denitrification methods happening in crop cultivation) (restrained relevant when soils under native conditions mean a large storage of carbon), which can be enumerated in terms of CO_2 counterparts.

2.6 *Influence of Climate on Perennial Crops*

Generally, perennial trees in moderate and cool subtropical temperatures lose their leaves and begin a cold toughening stage in early fall before becoming latent in late fall. In some plants, the termination of growth during fall is activated by shorter day length, while others respond to cooler air temperatures. Buds persist dormant or in the phase of 'rest' because of internal functional blocks (inhibitors) that stop their progress even under idyllic conditions for growth. These physiological blocks are detached when buds are exposed to chilling temperatures above freezing for some weeks. The chilling condition is often satisfied in winter for both high and low latitude species. Subsequent chilling is completed and buds are no longer in a state of 'rest,' they develop 'quiescent' and respond to heat accumulation (Fig. 2). Cold temperatures during the dormant period stop bud growth, whereas buds develop active losing much of their hardiness when the temperature becomes promising for growth. Perennial crops are mainly susceptible to cold damage at three distinct stages: 1. during the fall before the tree is sufficiently toughened, 2. during the winter latent period when severe cold measures can cause damage to woody tissue, and 3. during spring when temperatures somewhat below freezing may kill flower buds subsequent loss of cold toughness [31]. While cold damage in fall and winter can source stable damages to perennial trees, they are typically less recurrent and have a lesser result on year-to-year inconsistency in yield, associated with spring season freeze events. Production is also inclined by other climate factors, such as the amount of plant remaining moisture in the soil profile, circumstances during pollination, the incidence of lengthy and/or extreme heat or drought events, and the occurrence of hail during the growing season. Moreover, weather conditions, comprising temperature and humidity, contribute to the risk of insect pests and plant diseases, which also interrupt orchard and winegrower productivity. An initial step in weather estimate for perennial crops is the documentation of the serious growth stages and associated environmental factors for the crop in question.

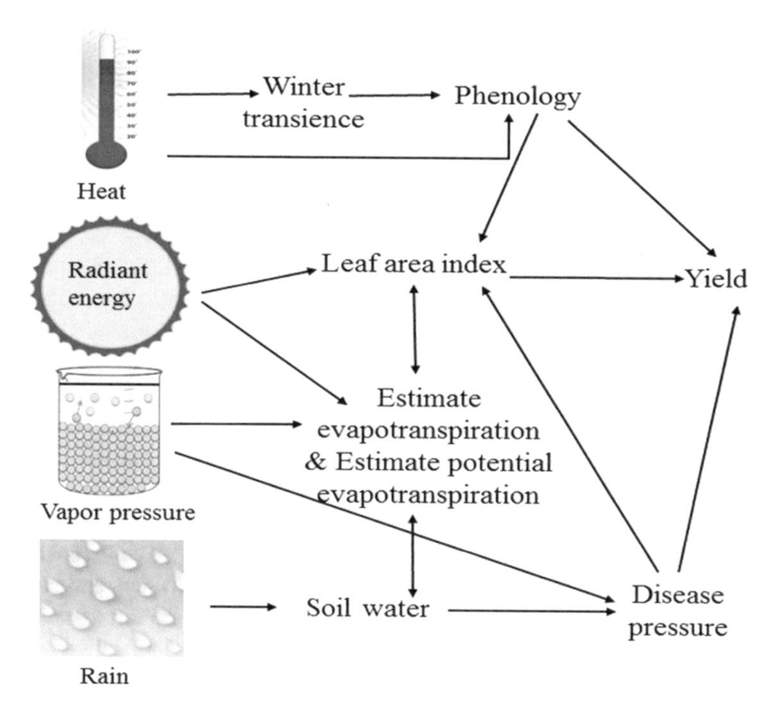

Fig. 2 Influence of weather and climate on perennial crop production and yield. The figure explains the influence of temperature variables on perennial crop cultivation [32]

2.7 *Local Impacts on Perennial Crop Production*

The inferior necessity of pesticide efforts, by evaluation with annual crops, signifies also an extra ecological benefit of perennial crops. Since, perennials take benefit of the usage of herbicides only during the planting stage of the crop, whereas annual crops need a year-round uses, and some energy crops, e.g., miscanthus and giant reed, exist no main illnesses necessitating plant protection measures. Useful properties are the lessening shares of chemicals ending up in soil, water, and air, causing injury to flora and fauna and disturbing human health. The rise of organic matter content in the soil results in soil carbon storage (carbon sequestration) and decrease of CO_2 in the atmosphere. Perennials (e.g., switchgrass, miscanthus, and giant reed) have a good ability to store carbon in the soil, largely due to their big and deep root expansion, since a large parcel of the organic carbon manufactured during photosynthesis rests in the ground in the postharvest. According to [33], carbon re-possession by the Miscanthus root and rhizome system is substantial, demonstrating c. 13.5–16.5 Mg ha^{-1} over the lifetime of the crop, and used carbon to soil from litter signifies c. 5.1–5.9 Mg ha^{-1} $year^{-1}$. In the classical repetition of cumulative carbon sequestration by switchgrass in the Mediterranean area, Virgilio et al. [34] computed an annual SOC buildup of 0.03–0.72 Mg ha^{-1} and a buildup of 5.5–7.2 Mg ha^{-1}

over 18 years. Furthermore, Virgilio et al. [34] also detailed that if switchgrass were grown in arable land, the subsequent secondary land use alteration effects were low when compared to the environmental benefits of the stored SOC. However, this soil–rhizome buildup can become harmful if land use changes, because of the discharge of the stored carbon. Perennial crop features (rapid growth, high yields, deep, and widespread root systems) describe the tolerance ability of these plants to minimal and contaminated soils. This aptitude offers the option of connecting soil decontamination and reinstatement with the production of biomass for biofuel and biomaterials with extra revenue. Also, use of marginal/contaminated soils pays to reduce the land versus food dilemma, the minimum direct and indirect negative effects due to land use change. The remediation volume of perennial grasses has been recognized by numerous studies, like those with Miscanthus, giant reed or switchgrass [35, 36]. Yet the competence between the crop and the particular marginal/contaminated soil to evade potential ecological and socioeconomic impacts should be taken into account. Sustainability of energy crop production in marginal/contaminated soils be determined by crop yields and crop's ability to return value to the land. Output loss in marginal/contaminated soils reduces the energy and greenhouse savings, but the presence of vegetation may pay to recover the quality of soil and waters and biological and landscape variety. The features of perennial crops also let the association of their cultivation with the use of wastewaters in irrigation. The application of treated wastewater to fields of perennials may pay to alleviate the shortage of water capitals and lessen the need for fertilizers, with global positive ecological outcomes. Irrigation with wastewaters might offer readily obtainable adequate amounts of N, P, and K and also adequate quantities of organic matter that recover the soil structure and other soil properties associated with the accessibility of water and nutrients. Yet the pros and cons of linking wastewater irrigation with perennial grass production should be sufficiently weighed, so that chances to produce sustainable biomass can be effective. Indeed, the presence of harmful substances in wastewaters can also be harmful to biomass growth and quality, and, if not precisely enclosed by the standup biomass, pollutants can collect in the soil or be leached to the ground and surface waters causing a threat to the bionetworks.

3 Financial and Socioeconomic Features of Perennial Crop Manufacture

3.1 Monetary and Financial Features of Perennial Crop Manufacture

Monetary examination is concerned with the extent of performance against set targets on every facet of a project. It recognizes the capability of use of resources and delivers ideas for refining the overall act. It also measures the efficiency of the organization in assembling the issues of production for the attainment of financial goals and supports the search for better methods. Lastly, it is a valuable tool for accountable areas of likely economic development, supporting management in their struggles toward the overall upgrading of performance. Economic analysis of biomass production includes three easily distinguishable steps. The first is farm income study, based on balance sheet and profit and loss accounts. This is based on an opening balance sheet and farm budgets sticking out income and expenditures for the subsequent years. The second step encompasses the approximation of future balance sheets based on farm sales and income predictions and on prospects concerning the timing of receipts and expenditures. This step classifies project connected future cash flows, which can be reached either straight (based on timed receipts from sales, etc., minus expenses for procurements and expenditures) or secondarily (based on net wages before devaluation plus changes in working capital). The third step is farm venture study. This uses cash flows from step 2 to approximate the attractiveness of the project by comparing future net inflows with the initial venture sum.

Monetary sustainability of perennial crops classifies onsets of financial feasibility pointers in contrast to another course of action for the supply of final products that may be shaped from bio-chains based on such industrial crops. From the lookout of the producer of bio-products (farmer, industry, dealer, depositor, etc.), adequate return to capitalized exertion or capital must be secured within reasonable risk levels, reasonably fast and with acceptable views for maintaining the activity in the forthcoming. With respect to the production of bioenergy, the European Commission has set high targets for carbon reduction and renewable energy influence to the EU energy sector. The targets for 2030 are much higher than the 2020 objectives and this signals a resulting growth of the cultivation of perennial energy crops. The possible value of the final products of perennial crops is restrained by the difference between the selling price and the projected annual equivalent life cycle cost, which is a degree of productivity. We accept no midway sales profit among the many actors along the bio-chains. Any positive overall profit margin is spread among all contributors (agricultural, conveyance, warehousing, transformation, marketing, etc.) rendering to relative influence and market forces.

3.1.1 Use of Minimal Land for Farming Industrial Crops

The European Commission has frequently professed the purpose to circumvent the cultivation of nonfood, and particularly energy crops in productive agricultural land, to avoid the substantial effects on food supply and charges. Direct or indirect land use variations, mostly initiated by renewable energy initiatives, have often affected the food market in many areas. So, it is not difficult to stimulate the cultivation of such perennial crops on various types of marginal land. Land rent varies significantly from region to region and the rent of marginal land is not set equal to zero, if the land has not possible for manufacture and income. For minimal land, a 40–70% discount off the payment of productive agricultural land was recognized, dependent upon the degree of nonconformity and other features. Nonetheless, as this is very much site and case specific, it is best to estimate the rent of marginal land at its opportunity cost, the income forgone due to the alteration of land use. Irrigation is another chief cost item, particularly for cultivation in marginal lands, since they are commonly water-stressed areas and the water may be transported from too far. Substantial amounts of energy and succeeding outlay may, therefore, be important for the irrigation of marginal lands. It has been noted still that the farming in marginal lands with minimal irrigation and other inputs is not typically an optimal choice, because of the excessively low agricultural yields. Endowments that may happen along with the bio-products chain should not be comprised of basic calculations. Such monetary inducements are best dignified after a basic assessment of the attraction of the project 'earlier grants' to display their effect distinctly.

4 Case Study on OPTIMA Project

4.1 Sustained Crops and Cost Chains

The OPTIMA project is devoted to the farming of perennial crops such as giant reed (*A. donax* L.), miscanthus (Miscanthus × Giganteus), and switchgrass (*Panicum virgatum* L.) in minimal nutritional soils in the Mediterranean region. The life cycle schemes comprise farming, yield and pretreatment, preparing and logistics, change, use phase, and end of life. Life cycle phase "farming" can be separated into the following processes: field groundwork, seeding/planting, upkeep including weed control, the application of fertilizer and irrigation, harvest, and clearing after a plantation's lifetime. Numerous issues are equal for each of the plants, counting the plantations' era of 16 years. However, the crop plant varies from each other with respect to the scale of inputs and outputs associated with the cultivation phase. Numerous value chains were analyzed within the OPTIMA project: (1) domestic heat, (2) collective heat and power (CHP), minor and large scale, (3) promoted pyrolysis oil, (4) biochar, (5) second-generation ethanol, and (6) 1,3-propanediol. These value chains form a characteristic mix of scales and claims that can both be obtained by the three perennial crops but also be suitable for Mediterranean countries.

4.2 Life Cycle Sustainability Assessment (LCSA)

A broadcast of LCSA for the cultivation and use of three designated perennial grasses: Miscanthus, giant reed, and switchgrass, was led as part of the sustainability assessment of the OPTIMA project. As a basic set-up, 21 groupings of three crops and seven use choices have been examined for their influences on seven environmental pointers in a screening LCA (Table 2).

Table 2 Details of perennial crops cultivation on the land with low agricultural values

S. no.	Factor	Component	Giant reed	Miscanthus	Switchgrass
1	Moistness elimination rate from the field	Proportion fresh matter	65	30	25
2	Elimination of biomass from the field	Dry matter/ha year	27.5	20	10.95
3	Irrigation	m^3/ha year	7000	7000	5000
4	Nitrogen fertilization	kg nitrogen/ha year	210	58	73
5	Phosphorous fertilization	kg phosphorous /ha year	80	37	37
6	Potassium fertilization	kg potassium /ha year	456	201	42

4.3 *Evaluation of Sustained Grasses and Application Options*

To deliver an overall situation of the result of the farming methods with respect to the adaptation and custom courses on weather change. It validates that the alterations in the results of the seven usage choices are larger than the changes between the biomass types. With respect to the position of crops and use options, it has to be underscored that some state situations such as desiccating and pelleting of all biomass meaningfully influence the results. Dependent on the case-specific environments, logistics chains could be calculated inversely. It gives an impression of the basic improvements in the OPTIMA project, all perennials and alteration/usage options scrutinized are exhibited.

4.4 *Environmental Impact Assessment*

In the setting of the OPTIMA project, the conservation related to the manufacture and use of perennial grasses grown on minimal lands in the Mediterranean region, concentrating on local impacts, were also dignified. Effects of producing these grasses on the biodiversity, scenery, soil quality, destruction, and use of water were measured and compared with the effects of nonrenewable fuels associated with idle land when bearing in mind only the cultivation phase. The study displayed that limited ecological impacts are mostly subjective to biomass cultivation. Miscanthus is the best performing crop at the local level since its low nutrient demand and high yield. Complete results recommended that these crops offer paybacks concerning soil properties and erodibility. Less tillage and high biomass production support biological and landscape diversity and impacts related to water resources can be lowered by adopting water- and energy-saving methods, a tactic particularly significant in the case of giant reed, which displayed the highest influence score to water use.

4.5 Cost-Effective Analysis

Present research within the agenda of the European Union (EU) project OPTIMA has exposed the monetary potentials of cultivation of perennial grasses in minimal lands of southern EU regions and their alteration into a number of final bio-products as part of a broader sustainability valuation. Three of the most likely perennial crops for Mediterranean weathers have been considered as raw materials for the manufacturing of a number of primarily bioenergy products: giant reed, M. × giganteus, and switchgrass. The study supposed disparate land marginality levels and associated the financial activities of the previous crops with the farming of the same plants in typical fruitful agricultural land. All crops need to be watered, at least in the initial years, to attain a good launch and positive growth. Cultivation in marginal land upsurges the need for agronomic inputs such as irrigation, fertilizers, etc., according to the lack of specific marginal land. So, from a monetary point of view, it was observed whether the difficulty of the augmented need for inputs is offset by the low economic rent of marginal land. In marginal lands within high rainfall regions, irrigation can be negligible and less costly (M0). In drier marginal regions, irrigation is vital for the existence of the plantation. Agronomic inputs in minimal land can be low (ML) or high (MH). In the typical productive agricultural land, the agronomic inputs are high (SH). Giant reed is more fruitful and in spite of higher agronomic expenditures, its cost per ton is lower (below the €75 ton^{-1} line in the high-input set-ups). On the other hand, Miscanthus is about as costly per ton of output as switchgrass (around €65–80), while it attains higher yields since it is more costly to grow than switchgrass. It is worth noticing that, because of lower marginal land yields, cultivation on low-quality land is more costly per ton of produced biomass, in spite of the lower chance rent of land and commonly minor amounts of agricultural inputs. Table 1 also displays the failure of costs by the main process category. It demonstrates the comparative implication of manuring, irrigation, and harvesting, making up about 50% of the normal annual cost of perennial grasses.

4.6 Socioeconomic Analysis

This unit debates the service properties in terms of total job counterparts per value chain using the reference units from the OPTIMA study. It also offers thorough figures for how these jobs are prearranged in terms of direct, indirect, and net-induced per value chain. As likely, the job requirement increases the amount of annually essential biomass supply increase. Giant reed-fueled value chains permanently exhibit a lower number of jobs as the crop yields at marginal land. In domestic heat, the total number of jobs equals is around 2–2.3%, with the number of direct jobs being in the range of 0.3%, and the net added induced jobs being around 2%. In small-scale CHP, direct jobs are equal to the induced ones, arising mostly from biomass production. In this value chain (as the previous one was moderately small and changes among crops were irrelevant), the influence of yields on the number of job equivalents shows the difference, in particular, for giant reed as it displays a much higher yielding volume (17.5 t ha^{-1} $year^{-1}$) than the other three, so less land is vital to secure fuel supply for the power plant with a subsequent lower number of direct job equivalents. The particular numbers of direct job counterparts range from 12 in the giant reed-fueled chain, to 15 for Miscanthus, and 22 for switchgrass.

4.6.1 Communal Sustainability

Communal sustainability approximates the influences of the value chain on society and rural progress. The study in OPTIMA took into account the following principles:

1. Provision to rural economy: Service is a major subject in rural economies. Certain value chains may persuade more regional job creation, inspiring the rural economy, while other value chains may be more focused on large-scale industry, regularly in the hands of international players/multinationals.
2. Local implanting: The capability of the local economy to progress and function a full value chain or part of it.
3. Nearness to markets: The pointer states the alteration between a more local method with low distances (feedstock transformed and expended locally) on the one side, and on the other side a more international/industrial method where the feedstock is conveyed to large industrial sites or to harbor areas to be dispersed.

All the perennial crops under study are dignified extremely valuable to the three social sustainability ethics as they are possible to expand agricultural activities, offer new openings for farmers and the rural economy, and enable a better frame for harvesting, storage, transport, and logistics. Table 3 validates the act of the under study value chains in the social sustainability criteria [37].

Domestic heat and small-scale CHP rank high in all three principles, as both the full value chain and the final product propose very good forecasts to the rural economy with the production of perennial crops, the manufacturing and/or increased

Table 3 Level of accomplishment of the under study value chains in social sustainability

S. No	Parameter	country side economy involvement	Confined implanting	Market vicinity
1	Inland heat	Strong	Strong	Strong
2	Combined heat and power (2 MWe small)	Strong	Strong	Strong
3	Combined heat and power (40 MWe large)	Moderate	Strong	Strong
4	Pyrolysis oil	Moderate	Low	Low
5	Biochar	Moderate	Low	Low
6	Bioethanol	Moderate	Low	Low

market for biomass boilers/related equipment, and the competence of service for their process and maintenance. Large-scale CHP value chains rank sensible in the effect to the local economy as they can be advantageous for the local economy in terms of partly providing the plant with raw material and creating jobs for building and functioning the plant. But the main part of biomass source and plant equipment is carried into the region from other regions or countries. The value chain ranks high in implanting to the local system as it can afford heat for district heating and electricity to the grid or manufacturing sites/businesses. The value chains of pyrolysis oil, biochar, and second-generation bioethanol rank low in surrounding to the local system and vicinity to markets. Because they are larger plants and the main part of their raw materials and particular sales of final product will be from outer the district/local economy.

5 Concluding Remarks

This chapter sketched the life cycle sustainability assessment and also the ecological, monetary, and socioeconomic features of perennial grass production and use, with a special focus on farming in Mediterranean marginal land. The assessed impact pathways depend principally on organization strength and crop traits, and second on the handling and use systems. Perennials can be measured as more ecologically appropriate crops than annual energy crops, since the prerequisite of inputs (fertilizers and pesticides) is low and the longer stability period in the ground aids erodibility, biodiversity, and use of resources. In addition, the replacement of fossil fuels and resources over the use of perennials can lead to energy investments and drops in greenhouse gas emissions, but regarding other ecological categories, e.g., acidifying emissions, negative impacts can be observed in association with conventional routes. When irrigation is compulsory, the impact on water resources is also harmful to the sustainability of the value chains. Their economical utilization is affected by yields and by the end use selections. Production in small-scale units brings paybacks toward job creation and the rural economy.

References

1. IEA India Energy Policy Review. 2020
2. Kumar CR, Majid MA (2020) Renewable energy for sustainable development in India: current status, future prospects, challenges, employment, and investment opportunities. Energ Sustain Soc 10:1–36
3. U.S Energy Information Administration. Country Analysis Executive Summary: India. Sep 30, 2020
4. Sumfleth B, Majer S, Thran D (2020) Recent developments in low iLUC policies and certification in the EU biobased economy. Sustainability 12:2–34
5. Guinee JB, Reijung R (1995) A proposal for the definition of resource equivalency factors for use in product life-cycle assessment. Environ Toxicol Chem 14:917–925

6. Klopffer W (2006) The Role of SETAC in the Development of LCA. Int J LCA 11 (3), May edition
7. Suski P, Speck M, Liedtke C (2021) Promoting sustainable consumption with LCA - A social practice based perspective. J Clean Prod 283:125234
8. Giesen CVD, Cucurachi S, Guinee J, Kramer GJ, Tukker A (2020) A critical view on the current application of LCA for new technologies and recommendations for improved practice. J Clean Prod 259:120904
9. Beemsterboer S, Baumann H, Wallbaum H (2020) Ways to get work done: a review and systematization of simplification practices in the LCA literature. Int J LCA 25:2154–2168
10. Moni SM, Mahmud R, High K, Dale MC (2019) Life cycle assessment of emerging technologies: a review. J Ind Ecol 24:52–63
11. Yadav P, Priyanka P, Kumar D, Yadav, Yadav K (2009) bioenergy crops: recent advances and future outlook. In Rastegari AA (eds), Prospects of renewable bioprocessing in future energy systems. Biofuel and Biorefinery Technologies 10:2019
12. McBride W, Greene C (2009) The profitability of organic soybean production. Renew Agric Food Syst 24:276–284
13. Scordia D, Scalici G, Brown JC, Robson P, Patane C, Cosentino SL (2020) Wild Miscanthus germplasm in a drought-affected area: physiology and agronomy appraisals. Agronomy 10:2–18
14. Fradj BS, Borzecka RM, Matyka M (2020) Miscanthus in the European bio-economy: a network analysis. Ind Crops Prod 148:112281.
15. Hu Y, Schafer G, Dupla J, Kuhn NJ (2018) Bioenergy crop induced changes in soil properties: A case study on Miscanthus fields in the Upper Rhine Region. PloS One 13:e0200901
16. Scordia D, Zanetti F, Varga SS, Alexopoulou E, Cavallaro V, Monti A, Copani V, Cosentino SL (2015) New insights into the propagation methods of switchgrass. Miscanthus Giant Reed Bioenergy Res 8:1480–1491
17. Mariscal MJ, Orgaz F, Villalobos FJ (2000) Radiation-use efficiency and dry matter partitioning of a young olive (*Olea europaea*) orchard. Tree Physiol 20:65–72
18. Pradhan S, Sehgal VK, Bandyopadhya KK, Panigrahi P, Parihar CM, Jat SL (2018) Radiation interception, extinction coefficient and use efficiency of wheat crop at various irrigation and nitrogen levels in a semi-arid location. Ind J plant physiol 23:416–425
19. Zhang J, Jiang H, Song X, Jin X, Zhang X (2018) The responses of plant leaf CO_2/H_2O exchange and water use efficiency to drought: a meta-analysis. Sustainability 10:2–13
20. Lewandowski SU (2006) Nitrogen, energy and land use efficiencies of Miscanthus, reed canary grass and triticale as determined by the boundary line approach. Agric Ecosys Environ 112:335–346
21. Smith R, Slater F (2010) The effects of organic and inorganic fertilizer applications to *Miscanthus × giganteus*, *Arundo donax* and *Phalaris arundinacea*, when grown as energy crops in Wales, UK. GCB Bioenergy 2:169–179
22. Copani V, Cosentino SL, Testa G, Scordia D (2013) Agamic propagation of giant reed (*Arundo donax* L.) in semi-arid Mediterranean environment. Ital J Agron 8:1–24
23. ISO, 2006a. ISO 14040:2006–Environmental management—life cycle assessment—principles and framework. International Organization for Standardization
24. ISO, 2006b. ISO 14044:2006–Environmental management—life cycle assessment—requirements and guidelines. International Organization for Standardization
25. Wu R, Yang D, Chen J (2014) Social life cycle assessment revisited. Sustainability 6:4200–4226
26. Schmidt WP, Sullivan JL (2002) Weighting in life cycle assessments in a global context. Int J LCA 7:5–10
27. Herron SA, Rubin MJ, Ciotir C, Crews TE, Van Tassel DL, Miller AJ (2020) Comparative analysis of early life stage traits in annual and perennial *Phaseolus* crops and their wild relatives. Front Plant Sci 11:34
28. Kulczycka J, Smol M (2016) Environmentally friendly pathways for the evaluation of investment projects using life cycle assessment (LCA) and life cycle cost analysis (LCCA). Clean Techn Environ Policy 18:829–842

29. Cilluffo A, Ruiz, NG (2019) World's population is projected to nearly stop growing by the end of the century. UN's "World Population Prospects 2019"
30. Finnan JM, Styles D (2013) Hemp: A more sustainable annual energy crop for climate and energy policy. Energy Policy 58:152–162
31. Burke MJ, Gusta LV, Quamme HA, Weiser CJ, Li PH (1976) Freezing and injury in plants. Annu Rev Plant Physiol 27:507–528
32. Winkler JA, Andresen JA, Guentchev G, Kriegel RD (2002) Possible impacts of projected temperature change on commercial fruit production in the Great Lakes region. J Great Lakes Res 28:608–642
33. Agostini F, Gregory AS, Richter GM (2015) Carbon sequestration by perennial energy crops: is the jury still out? Bioenergy Res 8:1057–1080
34. Virgilio ND, Facini O, Nocentini A, Nardino M, Rossi F, Monti A (2019) Four-year measurement of net ecosystem gas exchange of switchgrass in a Mediterranean climate after long-term arable land use. GCB Bioenergy 11:466–482
35. Arora K, Sharma S, Monti A (2016) Bio-remediation of Pb and Cd polluted soils by switchgrass: a case study in India. Int J Phytoremediation 18:704–709
36. Papazoglou EG (2007) *Arundo donax* L. Stress tolerance under irrigation with heavy metal aqueous solutions. Desalination 211:304–313
37. Keller H, Rettenmaier N, Reinhardt GA (2015) Integrated life cycle sustainability assessment—a practical approach applied to bio-refineries. Appl Energy 154:1072–1081

Sustainable Development: ICT, New Directions, and Strategies

Florin Dragan and Larisa Ivascu

Abstract This chapter aims to make an inventory of what sustainable development (SD) means. Evaluating the implications of the concept from 1951 to the present, one can observe a substantiation of the concept and an intensification of the involvement of organizations. The implications of individuals, organizations, local and international authorities are at different levels. From this perspective, it is important to conduct research on the implications, new directions developed, and strategies that can be developed based on individual and organizational characteristics, but also national and global conditions. This chapter evaluates the new activities of sustainability and the action of information and communication technology (ICT) on the 17 objectives of sustainability. At the same time, an evaluation of new technologies on life cycle sustainability assessment is performed. At the end of the chapter is presented a case study that highlights the importance of urban agriculture in the context of sustainable development.

Keywords ICT · Sustainability · Business · Machine learning · Artificial intelligence · Blockchain · Development · Social

---

F. Dragan
Department of Automation and Applied Informatics, Faculty of Automation and Computers, Politehnica University of Timisoara, Timisoara, Romania
e-mail: florin.dragan@upt.ro

L. Ivascu (✉)
Management Department, Faculty of Management in Production and Transportation, Politehnica University of Timisoara, Timisoara, Romania
e-mail: larisa.ivascu@upt.ro

Research Center for Engineering and Management, Politehnica University of Timisoara, Timisoara, Romania

S. S. Muthu (ed.), *Life Cycle Sustainability Assessment (LCSA)*,
Environmental Footprints and Eco-design of Products and Processes,
https://doi.org/10.1007/978-981-16-4562-4_5

1 Main Concepts

Sustainability is being addressed by more and more organizations. If in the 2000s the implications of European organizations [1, 2] were limited to environmental activities, now there is a complete involvement of organizations. Regardless of the field of activity, there is an involvement in the dimensions of sustainability: economic, social, and environmental. The banking field has an intense involvement in Corporate Social Responsibility (CSR). These implications can be divided into different categories: financial education, culture, sports, environment, human resources, and digitalization. The field of car construction [4] has an intense involvement in technological innovation and in reducing environmental pollution. The industry emphasizes the importance of improving the carbon footprint and supporting innovations involving blockchain, artificial intelligence (AI), Internet of Things (IoT), machine learning (ML), and other complex structures [3–5]. All these technologies contribute to achieving the 17 sustainability objectives and 169 targets. The field of tourism is involved in the support of the environment and the development of human relations. There are areas of activity that have reduced implications in sustainable development (trade, consulting, and others) [8]. Within these areas, the implications for sustainable development are limited to concerns for society and basic principles related to the environment (selective recycling and reuse). All organizations are involved in sustainability if they identify several benefits. These benefits differ depending on the position they hold individually in the stakeholder group [8–12]. The implications of organizations in sustainable development can be improved, but new directions of the concept and strategies that can be adopted to achieve several benefits targeted by organizations must be identified.

By definition, sustainability combines 2 important approaches:

a. *Intragenerational approach* (present needs that are met through the use of resources)
b. *Intergenerational approach (*not to compromise the ability of future generations to use resources and have the same rights).

This ethical dilemma requires a change in the economy that will contribute to the realization of other business models, production, and consumption.

The present chapter aims to structure a series of organizational strategies based on the new directions of the concept of sustainability. The chapter begins with the presentation of the evolution of the concept, then it continues with the presentation of the new business trends and ends with the structuring of some organizational strategies.

2 Evolution of the Concepts

In 1987, a definition was developed that has become a reference over time. It was drafted by Norwegian Prime Minister Gro Harlem Brundtland. The document was called "Our Common Future" [15]. This report was developed by the United Nations World Commission on Environment and Development. In view of this report, sustainability is defined as "the needs of the present without compromising the ability of future generations to meet their own needs" [13, 14, 23].

To identify the present approach to the concept of sustainability, it is necessary to carry out an analysis of the definitions and interpretations of the different perspectives [8–15]. Therefore, Table 1 presents such an analysis.

Evaluating the implications of the concept of sustainability in the following table, we can identify a series of common elements or some elements that outline the various approaches [18]. The concept of sustainability involves several connotations with reference to the environment, society, and economy. These connotations of sustainability refer to:

a. *Green*—The concept of green is identified with practices that are environmentally friendly, support natural resources, and the ability of future generations to have at least the same resources as today [16]. These practices are approached in more and more fields, and thus can be exemplified: green marketing, green product, green concept, green entrepreneur, and others. It is non-toxic, clean, organic, fair trade, ethical, artisanal, uses modern technologies, and streamlines the environmental footprint.
b. *Society*—Society is one of the important components of sustainable development because it refers to the living environment, quality of life, equal opportunities, poverty and other variables [18]. The society contributes to defining ecological boundaries and accelerating transitions to a sustainable world.
c. *Green Information technology*—Information technology (IT) influences sustainable development and has an important role in operations, processes, and organizational systems. The investments in IT made by the organizations have contributed to important innovations and equipment [17]. Green investments are associated with those investments made within organizations that have an impact on profit and protect the environment.
d. *Global action*—Global actions have been shaped by various scientific events that have taken place since 1951 and until today. Each event outlined a series of principles and directions that can be approached by organizations to increase capacity for sustainable development [19]. Currently, there is no obligation for organizations to be involved in sustainable development, but a number of aspects included in this concept become mandatory (for example Directive 2014/95 / EU).
e. *Education for Sustainability (EfS)*—Sustainability education must be initiated from primary schools when students learn certain concepts related to the environment. EfS emphasizes the role of teaching and learning for sustainability which is seen as a preferred condition. This condition implies a certain specific attitude and skills [21]. The development of an approach to learning from the first years of school contributes to the formation of individuals with solid principles. This learning has several forms: design education, project-based learning, survey-based learning, professional communities for learning, gamification, but

Table 1 The evolution of sustainability concept

No	Year	Activity	Short description	Elements identified in this activity
1	1951	"International Union for the Nature Conservation (UICN)"[1]	This is the first report that talks about the environment, at a global level, is published, focusing on achieving a balance between economy and ecology	Economy, ecology, environment
2	1969	"United Nations published the report entitled „Man and His Environment or U Thant Report"[2]	Approach to improving the environment. This report was prepared by 2,000 scientists	Environment
3	1970	Rome Club[3]	It outlines the possibility of depleting environmental resources	Resources
4	1972	"The first UN and UNEP World Conference on the Human Environment took place in Stockholm, Sweden"[4]	At this conference were published: an environmental action plan and a statement	Environmental conservation
5	1975	"UNESCO Conference on Environmental Education, Belgrade, Yugoslavia"[5]	The "Charter of Belgrade" statement set out a framework for environmental education	Education, environment
6	1975	"International Congress of the Human Environment (HESC), Kyoto, Japan"[6]	They pointed out problems like those identified in the 1972 Stockholm action	Environmental conservation
7	1979	"First edition of the World Climate Conference, Geneva, Switzerland"[7]	The need for research on climate change has been accentuated. It also aims to monitor programs	Climate change

(continued)

[1] https://www.iucn.org.

[2] https://sustainabledevelopment.un.org/rio20.

[3] https://sustainabledevelopment.un.org/rio20.

[4] https://sustainabledevelopment.un.org/milestones/humanenvironment#:~:text=The%20United%20Nations%20Conference%20on%20the%20Human%20Environment%20(also%20known,June%205-16%2C%201972.

[5] https://sustainabledevelopment.un.org/milestones/humanenvironment#:~:text=The%20United%20Nations%20Conference%20on%20the%20Human%20Environment%20(also%20known,June%205-16%2C%201972.

[6] https://sustainabledevelopment.un.org/rio20.

[7] https://sustainabledevelopment.un.org/rio20.

Table 1 (continued)

No	Year	Activity	Short description	Elements identified in this activity
8	1981	"The first UN conference on the least developed countries, Paris, France"[8]	During this conference, steps are being taken to help developing countries. A report has been developed which includes various guidelines and measures	Priorities for underdeveloped countries
9	1984	"United Nations World Commission on Environment and Development—WCED"[9]	WCED adopts comprehensive development plans for environmental conservation and is concerned with developing cooperation between countries	Environmental conservation
10	1987	The WCED Our Common Future report has been published (Our Common Future)[10]	In this meeting, the concept and the associated principles are substantiated	Principles for sustainable development
11	1987	The Montreal Protocol has been published[11]	It outlines the harmful effects on the ozone layer	Harmful effects on the environment
12	1990	"Second World Climate Conference, Geneva, Switzerland"[12]	It aims to support the steps for climate change and the development of a system to carry out a monitoring activity of this impact	Climate change
13	1992	"United Nations Conference on Environment and Development, Rio de Janeiro, Brazil (United Nations Conference on Environment and Development)"[13]	All the principles and actions of sustainable development are set out in Agenda 21	Principles set for sustainable development (economic, social, and environment)

(continued)

[8] https://sustainabledevelopment.un.org/rio20.

[9] https://sustainabledevelopment.un.org/rio20.

[10] https://sustainabledevelopment.un.org/content/documents/5987our-common-future.pdf.

[11] https://sustainabledevelopment.un.org/content/documents/5987our-common-future.pdf.

[12] https://unfccc.int/resource/ccsites/senegal/fact/fs221.htm.

[13] https://www.un.org/en/conferences/environment/rio1992.

Table 1 (continued)

No	Year	Activity	Short description	Elements identified in this activity
14	1997	"Kyoto Climate Change Conference, Kyoto, Japan"[14]	This protocol aims to reduce the amount of Co2 and GHG. It has been in force since 2005 and is called the Kyoto Protocol	Greenhouse gas emissions
15	2000	The UN has published the Millennium Declaration[15]	The eight Millennium Development Goals are set to be reached by 2015	Environment, Economic, Social
16	2002	"World Summit on Sustainable Development, Johannesburg, South Africa"[16]	It establishes and consolidates previous obligations that sustain the steps of sustainability	Future directions (green, information technology)
17	2009	Third World Climate Conference, Geneva, Switzerland[17]	It aims to anticipate disasters and strengthen the climate change monitoring system	New directions for climate change
18	2009	G20 World Congress Summit, Pittsburgh, United States[18]	The moderate and sustainable economy is a direction targeted by the G20 member states	Moderate economy and economy that integrates the principles of sustainability
19	2012	"UN Conference Rio + 20, Rio de Janeiro, Brazil (UN conference Rio + 20)"[19]	Twenty years after the Rio conference, the Our Common Future report renews its commitment to the goals of sustainable development (GSDs) and encourages the problems of the green global economy	Green global economy, GSDs
20	2015	UN Summit on Sustainable Development 2015, New York, SAD[20]	This event establishes the 17 Millennium Development Goals. They must be achieved by 2030. These directions are presented in the UN Agenda 2030	17 Millennium Development Goals

(continued)

[14] https://www.un.org/press/en/1997/19971201.ENV453.html.

[15] https://sustainabledevelopment.un.org/rio20.

[16] https://sustainabledevelopment.un.org/milesstones/wssd.

[17] https://sustainabledevelopment.un.org/rio20.

[18] https://sustainabledevelopment.un.org/rio20.

[19] https://sustainabledevelopment.un.org/rio20.

[20] https://sustainabledevelopment.un.org/post2015/summit.

Table 1 (continued)

No	Year	Activity	Short description	Elements identified in this activity
21	2015	UN Conference on Climate Change COP 21, Paris, France[21]	Agreement to reduce greenhouse gases to limit global warming	Greenhouse gases
22	2016	"UN Conference on Climate Change COP 22, Marrakech, Morocco"[22]	This action aims to reduce GHG emissions, water management by improving water scarcity, water sustainability in all countries (including those where resources are limited). It aims to use energy sources that generate a low amount of carbon	Water management, Waste management, GHG management
23	2017	"G20 summit in Hamburg, Germany"[23]	The theme of the event was "Modeling an interconnected world"	Interconnected solutions
24	2018	"UN Conference on Climate Change COP 24, Katowice, Poland"[24]	More than 23,000 delegations from 190 countries met at three major events included in sustainability concerns	Climate Change, GHG, national actions
25	2019	"UN Climate Change Conference (UNFCCC COP 25), 2–13 December 2019, Madrid, Spain"[25]	This action targeted several directions on climate change and new directions	Climate change, global actions
26	2020	"UN Climate Change Conference (UNFCCC COP 26), 9–19 November 2020, Glasgow, United Kingdom"[26]	Carried out in pandemic conditions in which the challenges of climate change were addressed	Covid-19 impact on sustainable development

[21] https://unfccc.int/process-and-meetings/the-paris-agreement/the-paris-agreement/key-aspects-of-the-paris-agreement.

[22] https://enb.iisd.org/climate/cop22/.

[23] https://www.g20.org.

[24] https://cop24.gov.pl.

[25] https://unfccc.int/cop25.

[26] https://unfccc.int/process-and-meetings/conferences/glasgow-climate-change-conference.

also other forms. EfS offers students and teachers an important mission, that of developing a clean, sustainable future in conjunction with the development of systemic thinking and design [20].

f. *Waste management*—The development of industries and globalization have led to an increase in the amount of waste. In this situation, they are added to household consumers who generate waste. The amount of waste started to increase and because of the improvement of the quality of life and the consumption behavior [23]. Therefore, an efficient waste management is an essential condition in the present society. The European Union is taking steps and developing directives to increase national capacity for recycling and reduce waste incineration. The recycling rate of the European Union is about 60% by 2020, and in many of the component countries, the recycling rate does not reach 20%. Therefore, measures are needed to develop an efficient and integrated waste management at the level of each EU Member State.
g. *Water management*—The amount of water used in nutrition, but also in the residential sector is a challenge for sustainability. The development of the business environment and the production capacity have led to an increase in the amount of water used. Wastewater recycling measures are being reduced in many countries around the world [21]. Being a natural resource, it must be used in a balanced way so as not to endanger the well-being of future generations.
h. *Green global economy (GGC)*—aims at a green economy that is defined by a low amount of carbon dioxide, the efficient use of resources, and the involvement of individuals or society. The benefits of a green economy are registered at the organization, but also for the society [22]. There are a number of approaches that measure GGC performance for different countries.
i. *Greenhouse gas emission*—carbon dioxide emissions but also the rest of GHG are the drivers for climate change and changes that take place in the environment. Carbon dioxide is an important gas, with a share of about 65% of GHG [23]. It reaches the atmosphere by burning fossil fuels, municipal and solid waste, certain chemical activities, and burning wood products. It can be eliminated by plants through their biological process (as a natural activity). The rest of the gases are released during agricultural activities, animals, municipal landfills, burning fossil fuels, industrial processes, and others. These gases deplete the ozone layer and can intensify the potential for global warming.

It can be seen that sustainability has a number of implications that are challenging for organizations and develops a number of organizational opportunities. Organizational involvement depends on organizational culture, shareholders, and financial capacity. Below are presented real practices of sustainability, but also the types of organizational reporting.

3 True Business Sustainability

3.1 Sustainability Reporting

Sustainability reporting is a voluntary activity, financial or non-financial, for companies and can be done annually or biannually, most of the time. More and more companies are reporting their sustainability as a condition of the business environment. This sustainability reporting involves evaluating the company's activities on the three dimensions of sustainability: economic, social, and environmental. The organizations evaluate the involvement in sustainability to highlight the main activities carried out since the last report. Recently, there has been an intensification of the companies' implications in reporting sustainability. At the global level, there is an approach called the Global Reporting Initiative (GRI) [8, 11, 23, 24]. This approach is structured on the three dimensions of sustainability and aims to consolidate a tool that can be used for each reporting, and the targeted indicators can be compared from one reporting to another.

Non-financial reporting targeting social and environmental indicators is gaining importance in the business environment.

- Qualitative reporting—the data provided must be coherent, relevant, current, and understood by stakeholders.
- Quantitative reporting—is based on different performance indicators that can be established for each field of activity.
  Semi-quantitative reporting—a combination of qualitative and quantitative reporting.

3.2 New Global Approaches to Sustainability

New approaches to sustainability address a number of issues related to information technology and environmental implications.

The implications of the 17 objectives of sustainable development (17 SDGs) are presented in Table 2. The names of the objectives are set by the 2030 Agenda in 2015 and are adopted by all United Nations Member States [23].

A. **Urban agriculture (UA)**—this concept involves the cultivation, processing, and distribution of food in or near urban areas. There is a growing tendency for this concept especially in these pandemic areas of 2020–2021. A case study is presented in the next chapter. There are a number of practices in the world that are presented below. It contributes to transforming the living environment into an eco-friendly and sustainable one. The benefits of urban agriculture are identified in energy, reducing costs with agricultural management, saving water and land, obtaining organic products, and developing well-being. This concept also contributes to improving the health of communities by involving residents

Table 2 The 17 objectives of sustainable development (17 SDGs)

Objective	Acronym	Objective	Acronym
No poverty	SDG 1	Reduces inequalities	SDG 10
Zero hunger	SDG 2	Sustainable cities & communities	SDG 11
Good health and well-being	SDG 3	Responsible consumption and production	SDG 12
Quality education	SDG 4	Climate action	SDG 13
Gender equality	SDG 5	Life below water	SDG 14
Clean water and sanitation	SDG 6	Life on land	SDG 15
Affordable and clean energy	SDG 7	Peace, justice and strong institutions	SDG 16
Decent work and economic growth	SDG 8	The power of partnerships	SDG 17
Industry, Innovation, and Infrastructure	SDG 9		

in recreational and ecological work. UA can improve social capital and civic engagement [15].

In France, there are businesses that grow strawberries in old containers that are being modernized. These containers are upgraded with light and aeroponics systems. Another practice is that of urban farmers which are formed in the form of a network. This network has the mission to promote these practices in other areas as well [17].

In Belgium, buildings integrate aquaponics to reduce the impact on the environment. Various steps have been taken to create small spaces for growing vegetables in urban areas.

The United States has implications for urban agriculture. Here are a number of mobile platforms that connect farmers with restaurants and consumers. Information platforms are being developed to change the behavior of supermarket consumers with the production of vegetables in their own gardens [19].

Germany and Austria have a number of implications for urban agriculture. There is a tendency to intensify the use of urban agriculture for own consumption, but also as a method of recreation for stressed and tired employees [20].

In Malaysia, this concept is appreciated, and the inhabitants are involved and support the principles of this concept. The results obtained are presented below [7, 20].

In Romania, there is no intense involvement, but the inhabitants know this concept, but they were not motivated to get involved. The results obtained are presented in the next section [9].

Urban agriculture contributes to achieving some objectives of sustainable development [23]. The situation is presented in Table 3.

Table 3 The implications of urban agriculture on the SDGs

No	Acronym	The implications of urban agriculture on the SDGs
2	SDG 2	The obtained products can be used for own consumption, but also for marketing
3	SDG 3	Cultivated fruits and vegetables contribute to the well-being of individuals and to the benefit of mental health
10	SDG 10	AU reduces economic disparities and contributes to improving access to organic products
11	SDG 11	The AU contributes to the development of a food resilience system
13	SDG 13	If AU practices are implemented correctly, they can lead to improved human impact on the climate

B. Information and communication technology (ICT)—provides a range of tools that contribute to the generation, processing, evaluation, and analysis of organizational data, improving global communication and streamlining organizational processes.

From the ICT perspective there are 2 dilemmas:

a. ICT contributes to improving environmental conditions and contributes to reducing pollution, streamlining processes, innovation and developing a sustainable approach.
b. ICT adds to the consumption of organizational resources, and thus contributes to consolidating unsustainable behavior.

In order to have an answer to these dilemmas, the involvement of the organization and the evaluation of the consumption of resources should be evaluated in order to establish the imprint on the environment. Such an investigation could answer whether ICT contributes to the development of sustainable or unsustainable behavior [9, 13]. An assessment of the implications of ICT on the three dimensions of sustainability would be the following:

a. Social: What is the contribution of ICT to social development? Can it support local communities to reduce unsustainable impact? How can the thinking and education of individuals be improved by using ICT to develop a sustainable society?
b. Economic: How can ICT change the structural component of the production economy so as to develop sustainable and innovative processes?
c. Environment: How can ICT reduce the negative impact of industries and the residential sector on the environment? What are the actions that need to be implemented to use ICT in improving environmental conditions?
   To answer this question, an evaluation of the benefits of ICT on the 17 SDGs [23] is presented in Table 4.

III. Innovation—helps streamline and optimize organizational processes. Innovation has a positive impact on organizational performance A number of areas have registered sustainable innovations as follows: public electric transport,

Table 4 The implications of ICT on the SDGs

No	Acronym	The implications of ICT on the SDGs
1	SDG 1	More than 2 billion people do not own a banking product. ICT provides access to information and can combat poverty
2	SDG 2	ICT can improve the life of the individual or organization. Provides access to connect
3	SDG 3	Computer health, telemedicine, direct interaction with patients are the facilities of ICT
4	SDG 4	The International Labor Organization is working to provide access to digitalization for millions of people
5	SDG 5	ICT contributes to providing digital access for women and men alike
6	SDG 6	Smart water and sanitation management are two approaches to ICT
7	SDG 7	ICT contributes to the development of green energy
8	SDG 8	Digital transformation contributes to the development of sustainable jobs and a sustainable economy
9	SDG 9	With the advancement of technology, new complex structures are being developed that can be approached organizationally for innovation and infrastructure improvement
10	SDG 10	It reduces inequality between countries and provides equal access to information
11	SDG 11	Offer the transition to smart cities and green communities
12	SDG 12	The development of ICT-based consumption strategies contributes to behavior modeling
13	SDG 13	Develops national and international policies for sustainable actions
14	SDG 14	Radio frequency and other facilities provide underwater life monitoring
15	SDG 15	Improves life on land through technological facilities
16	SDG 16	Contributes to the development of indicators and facility for improvement
17	SDG 17	Connecting communities and networks is done through ICT

electric trucks, cheap energy storage, long-term storage, plastic recycling, efficient lighting, accessible solar power, carbon management, innovative learning methods, and others. [16, 23].

An assessment of the main implications of innovation on sustainable development [23] is presented in Table 5.

IV .Artificial Intelligence—is defined by the Encyclopedia Britannica, as "the ability of a digital computer or computer-controlled robot to perform tasks commonly associated with intelligent beings". So, AI is an entity that can receive input, interpret it, learn, and exhibit behaviors that help the entity achieve a certain goal [17, 23].

Table 6 presents the implications of Artificial Intelligent on the SDGs [23]. The implications of AI on the 17 SDGs are presented.

Table 5 The implications of Innovation on the SDGs

No	Acronym	The implications of innovation on the SDGs
3	SDG 3	Medical innovations contribute to improving the health of the population
4	SDG 4	Educational innovations develop new competencies and educational approaches
6	SDG 6	New innovations in water management are developing new approaches
7	SDG 7	Green energy innovations offer repositioning opportunities
9	SDG 9	Great innovations in industries and infrastructure contribute to the development of sustainable societies and economies
11	SDG 11	Investing in city innovations contributes to the development of smart community networks
13	SDG 13	Globally, there are concerns about the innovative approach to climate change

E .Machine Learning—the development of new materials for the realization of some applications for the society is a preoccupation of artificial intelligence. Machine learning helps people interpret and make sense of complex data. Experiments produce a very large amount of data, and the individual is deposited by this amount. In this sense, machine learning gives meaning to these data in order to make organizational decisions agreed by the concept of sustainability. We can also talk about science-aware machine learning or physics-informed machine learning [1, 6, 9].

F .Blockchain—has an important role in sustainable development. It helps to encourage collaboration between consumers and producers. Consumers are helped to identify healthy and sustainable lifestyles. Manufacturers are helped to improve their supply practices and waste management system. Blockchain helps identify buyers, how much to buy, how to buy, and where to buy. More and more buyers want sustainable products that contribute to the development of sustainable behaviors [7, 16, 23].
Table 7 presents the implication of blockchain to the 17 SDGs [23].

G .Internet of Things—It provides support for connecting multiple machines. They collect real-time data to help solve specific problems. Many organizations use IoT to make energy efficient, clean energy, and build responsible and sustainable behavior. IoT contributes to the measurement and remote control of certain things. In the past, this was not possible because there is no advance in technology [14, 21].
Table 7 presents the principal implications of IoT on the 17 SDGs [23]. The major implications of IoT in sustainable development are briefly presented (Table 8).

H .Complementary structures—other complementary structures offer different opportunities for sustainable development. Here are various innovative approaches that contribute to the efficiency and improvement of organizational activities. The large amount of data implies new informational approaches because the individual is dependent on the processing capacity and the decision-making capacity [23].

Table 6 The implications of AI on the SDGs

No	Acronym	The implications of ICT on the SDGs
1	SDG 1	Using big data analytics to map poverty. Using the concept of sharing economy as a business model. Reducing the routine at work
2	SDG 2	It contributes to the improvement of the nutritional level of the population. The use of satellite images helps to improve the conditions in certain areas. Soil assessment reduces the negative effects of nutrition. The predictions generated can contribute to agricultural planning
3	SDG 3	Support for the development of applications that contribute to improving the health of the population. The concepts are met: preventive healthcare, predictive healthcare, cognitive healthcare, and personalization of treatment schemes
4	SDG 4	Support for the development of financial and educational applications. The following concepts are outlined: interactive massive open online course (MOOC), challenge-based learning (CBL), global classroom, and personalized learning programs based on avatars
5	SDG 5	Monitoring gender equality in different countries and in different industries through the use of maps. Development of innovative technologies for the development of women's skills
6	SDG 6	The smart water monitoring approach can be outlined
7	SDG 7	The smart renewable energy grinds approach can be outlined
8	SDG 8	Reduces inhuman jobs and contributes to the development of new products and services in line with economic growth
9	SDG 9	Planning, the inclusion of robots, expert systems and other facilities are used to improve organizational activity. The need for smart factories is accentuated
10	SDG 10	Contribution to the development of new life chances for people with disabilities (for example prosthesis for tomorrow and legs)
11	SDG 11	The following can be outlined: smart sensors to identify congestion in certain areas or other details. Smart cities involve the AI approach
12	SDG 12	Recognizing images or identifying consumption behaviors helps sustainable development (optimal production level)
13	SDG 13	Strategies are being developed for better air pollution management, with an emphasis on renewable energy capacity
14	SDG 14	Helps reduce production emissions and carbon footprint. It also aims to predict disasters
15	SDG 15	AI protects the basic species in the water. It also monitors illegal fishing activities and evaluates water exploitation
16	SDG 16	Monitors: population trends, epidemics, the development of some diseases and others
17	SDG 17	Develops responsible and transparent services that contribute to the development of smart governance

Table 7 The implication of blockchain to the SDGs

No	Acronym	The implications of blockchain on the SDGs
1	SDG 1	Support in accessing information
2	SDG 2	It offers the traceability of the production system and of life
3	SDG 3	Data security and crowdsourcing health systems
4	SDG 4	Innovative approaches in learning systems
5	SDG 5	Structures developed in a balanced and transparent way
6	SDG 6	Internet of Things Management Systems
7	SDG 7	Supports transparency of local markets and transactions of various types
8	SDG 8	It offers directions for optimizing financial markets and business models for economic growth
9	SDG 9	Elimination of intermediaries and development of structures according to organizational culture
10	SDG 10	Supports: sharing economy, and financial access
11	SDG 11	Supports: collective decisions, concept of smart cities and identification of market characteristics
12	SDG 12	Ensures the traceability of production processes
13	SDG 13	Ensures transparency of reducing the amount of greenhouse gases
14	SDG 14	Supports collaboration and monitoring of activities
15	SDG 15	Supports collaboration and monitoring of entities from different areas
16	SDG 16	Supports transparency and democracy in public structures
17	SDG 17	Supports global cooperation

3.3 *The Importance of ICT Within the LCSA*

This section presents an assessment of the opportunities offered by ICT for LCSA. Innovation and IT support is important for every stage of the life of a product or service. From this perspective, this chapter makes an inventory of the power of ICT to reduce the impact of a product or service on the environment throughout its life.

Main stage in LCSA	ICT implications
Design	The AI used in product design provides computer support for performing various process activities. Machine learning is important for product research and development. IoT streamlines human–computer interaction and contributes to cost efficiency. ICT plays an important role at this stage
Extraction of raw materials	The extraction of materials requires a high degree of innovation and intense use of information technology. At this stage, AI, IoT, machine learning and other implications can be used

(continued)

(continued)

Main stage in LCSA	ICT implications
Manufacturing	Manufacturing processes include in each ICT activity and their facilities. IoT is used to monitor the speed of production equipment. The employed staff can observe the details of the production activities and many other activities. ML offers producers the opportunity to increase production volume, accelerate research development, and bring improvements to the supply chain. Blockchain can provide data on identity management, product quality, source of materials, monitoring, and more. The complementary structures of ICT contribute to the efficiency of the production activity
Packaging	Packaging is a priority for many industries because waste management involves many activities for optimization. From this perspective, new IT concepts must be considered to improve the packaging process. Here we can mention IoT which requires the placement of a bar code for easy identification of products. AI can be used to capitalize on customer data and a continuous change of products to suit the market
Distribution	The use of AI in distribution contributes to the improvement of products through the stimulation and experience in continuous progress offered by an automatic feedback system. At the same time, AI can accompany each stage of the distribution to reduce the impact on the environment. IoT connects vehicles, pallets, locations, and other entities. Data can be used efficiently to improve the environmental footprint of products
Product use	The use of ICT in the use of products must be balanced so as not to balance between the two dilemmas (ICT contributes to the development of its sustainability and can it have a negative effect?). At this stage, AI, IoT, machine learning, and other complementary structures are used
End of life	Warranty certificates can be monitored using IoT. At the end of life, products should follow one of the functions of sustainability to reduce the carbon footprint. Therefore, ICT contributes to the consolidation: recycling, reuse, reconditioning, renovation, repair, remanufacturing, redesign, reimaging, reduction or recovery. The disposal of products should be reduced from the perspective of sustainable development

4 Case Studies

This section presents the importance of urban agriculture in new approaches to sustainable development. The pandemic period and the changing preferences of individuals contribute to strengthening the desire for urban agriculture. This concept is found both at the household level and in organizations that emphasize the need to develop sustainable jobs.

Table 8 The implication of IoT on the SDGs

No	Acronym	The implications of IoT on the SDGs
1	SDG 1	Allows access to information to all individuals
2	SDG 2	It contributes to the improvement of the annual production of cereals in order to reach the entire population. Producing more grain and reducing waste are principles supported by the IoT. IoT-based systems used for land irrigation optimize water consumption and reduce waste
3	SDG 3	It contributes to improving the rate of access to basic facilities to develop a healthy life
4	SDG 4	Using facilities to increase the quality of education
5	SDG 5	Supports transparency for all developed applications
6	SDG 6	Contributes to the support provided to entities with difficulties in water supply
7	SDG 7	Provides access to clean energy to all people
8	SDG 8	The development of tools that include IoT contributes to economic growth and eliminates abusive work
9	SDG 9	It has a big role in industry, innovation and infrastructure
10	SDG 10	Supports solutions to reduce inequalities and provide opportunities for all
11	SDG 11	IoT input is increasing for sustainable communities
12	SDG 12	IoT offers solutions for optimizing consumption and developing responsible behaviors
13	SDG 13	The use of IoT in climate change management contributes to the development of opportunities to reduce the amount of greenhouse gases generated
14	SDG 14	The use of IoT in shipping, which is the most widely used globally, can reduce the amount of fuel used by 15%
15	SDG 15	It involves connecting between different entities and reduces interaction barriers
16	SDG 16	Supports transparency and democracy in public structures
17	SDG 17	Supports global cooperation and partnership

4.1 Case Study Details

This case study presents the perceptions of the inhabitants of Romania and Malaysia on urban agriculture. A questionnaire was applied using the online environment to identify people's perceptions in different directions. This research took place in the period 2020–2021. 300 respondents for each country responded. This is an exploratory study. Below is a selection of the data obtained.

4.2 Results—A Results Section

Below are presented a series of concepts of urban agriculture and the results recorded by respondents, Table 9.

Table 9 Selection of answers from the research

Direction	Romania	Malaysia
Knowledge of the concept of Indoor UA	54% of respondents know the concept and were involved	75% of respondents know the concept and were involved
Knowledge of the concept of Outdoor UA	75% of respondents know the concept and were involved	100% of the respondents know the concept and were involved
Importance of UA for sustainability	60% of respondents positively appreciate the importance of AU for SD	95% of the respondents positively appreciate the importance of AU for SD
It is difficult to start UA	70% of respondents consider it difficult	15% of respondents consider it difficult
Financial involvement in the UA	53% of respondents want to get financially involved in the AU	35% of respondents want to get financially involved in the AU
I live in a community that practices UA	28% of respondents live in the AU community	91% of respondents live in the AU community
The family is educated for the UA	34% of respondents answered in the affirmative	94% of respondents answered in the affirmative
Your friends are educated in the direction of the UA	54% of respondents say they are educated in the direction of AU	74% of respondents say they are educated in the direction of AU
You feel responsible for practicing UA	45% feel responsible	83% feel responsible
My university shares the concepts of UA	30% of respondents say that the university shares concepts of AU	90% of respondents say that the university shares concepts of AU
The UA contributes to the dirtying the place the place where it is practiced	85% of respondents say that the place where AU is practiced generates dirt	38% of respondents say that the place where AU is practiced generates dirt
Indoor workplace UA is time consuming	55% agree with this statement	28% agree with this statement
UA generates an odor that bothers me	61% agree with this statement	35% agree with this statement
I am ready to adopt the UA in the future	70% agree with this statement	95% agree with this statement
I am ready to invest in AI for good results	45% agree with this statement	35% agree with this statement

4.3 Discussion

Evaluating the results obtained, it can be seen that Romania, which is a developing country, has a limitation of involvement in urban agriculture, especially when it requires financial involvement. The culture for urban development exists in Romania and Malaysia. In Malaysia, the culture for reducing the impact on the environment and involvement in urban activities is a consolidated one. Respondents in Malaysia appreciate urban beekeeping and are willing to get involved in consolidating activities in the future. Respondents in Romania have a lower desire to get involved in urban agriculture, but the percentage is significant. Romanians are willing to invest in information technology to facilitate their activity and increase their innovation capacity. From this aspect, the innovation needs that exist at the national level can be highlighted. In contrast, Malaysia does not identify a major need for investment in IT, as evidenced by the lack of innovation found in Malaysia.

If we evaluate the implications of universities in sharing the concepts related to AU, it can be seen that in Romania, only 30% of respondents appreciate that there is an involvement of universities. In contrast, in Malaysia, 90% of universities support and share knowledge about urban agriculture.

The Romanian communities have a low involvement in the AU, as the respondent appreciates. In Malaysia, communities are intensely involved and support the concepts of sustainability.

Therefore, this concept is of interest and must be strengthened at the individual and organizational level.

5 New Directions and Strategies

New strategic directions can be developed on the three dimensions of sustainability: economic, social, environmental. First of all, these approaches aim at involving information technology in the proposed and implemented activities.

The first strategic directive must aim at **sustainability management**. Thus, all organizational operations and processes must be evaluated in order to develop a new organizational culture. The commitment of all parties involved in the decision-making process must contribute to achieving a sustainable reputation. The management team has the role of mediator and must support a balanced involvement of all employees.

Risk management is a process that must be addressed in every organization. Regardless of whether certain risk management activities are outsourced or are performed within the organization, they must be performed periodically. Risk management is important for the sustainable development of the organization.

Sustainable innovation must be considered for the adoption of the most efficient sin and solutions that integrate information technology. Innovations help organizations reduce their environmental footprint and improve their organizational impact on the environment. These activities can be achieved through efficient and integrated management. This directive must emphasize the nature of **eco-efficiency**.

Sustainable competitive advantage is an opportunity for organizations aiming for sustainable development. Through a good positioning on the market, innovative investments, the development of solid partnerships, the consolidation of the market share, and other indicators contribute to the development of the competitive advantage.

6 Conclusions

Sustainability is approached globally, being an appreciated approach. The involvement of organizations in sustainable development is voluntary and generates a number of benefits to the organization and society. Strengthening the organizational capacity for sustainable development can be achieved through consolidated activities on the three dimensions: economic, social, and environmental.

This chapter presents the concept of sustainability, since its first appearance. It continues with the realization of an evolution of the concept of sustainability for which an identification of the key concepts encountered throughout the evolution is also achieved. Throughout the evolution, it can be seen that information technology has an important role in substantiating the role of sustainability in organizations.

A case study is also presented in which urban agriculture is evaluated. UA is constituted as an increasingly approached concept and of interest to individuals. This case study was conducted in Romania and Malaysia. The case study is constituted in the form of an exploratory research. At the end of the chapter are presented new strategic directions for organizations, based on the requirements of the business environment.

References

1. Abduljabbar R, Dia H, Liyanage S, Bagloee SA (2019) Applications of artificial intelligence in transport: an overview. Sustainability (Switzerland). https://doi.org/10.3390/su11010189
2. Andreopoulou ZS (2013) Green informatics: ICT for green and sustainability. J Agric Inform. https://doi.org/10.17700/jai.2012.3.2.89
3. Añón Higón D, Gholami R, Shirazi F (2017) ICT and environmental sustainability: a global perspective. Telemat Inf. https://doi.org/10.1016/j.tele.2017.01.001
4. Bai C, Sarkis J (2020) A supply chain transparency and sustainability technology appraisal model for blockchain technology. Int J Prod Res. https://doi.org/10.1080/00207543.2019.1708989
5. Brown DG, Verburg PH, Pontius RG, Lange MD (2013) Opportunities to improve impact, integration, and evaluation of land change models. Curr Opin Environ Sustain. https://doi.org/10.1016/j.cosust.2013.07.012
6. Çınar ZM, Abdussalam Nuhu A, Zeeshan Q, Korhan O, Asmael M, Safaei B (2020) Machine learning in predictive maintenance towards sustainable smart manufacturing in industry 4.0. Sustain 12(19):8211. https://doi.org/10.3390/su12198211
7. Cioffi R, Travaglioni M, Piscitelli G, Petrillo A, De Felice F (2020) Artificial intelligence and machine learning applications in smart production: Progress, trends, and directions. Sustainability (Switzerland). https://doi.org/10.3390/su12020492
8. Di Vaio A, Boccia F, Landriani L, Palladino R (2020) Artificial intelligence in the agri-food system: Rethinking sustainable business models in the COVID-19 scenario. Sustainability (Switzerland). https://doi.org/10.3390/SU12124851
9. El-Haggar S, Samaha A (2019) Sustainability. In: Advances in science, technology and innovation. https://doi.org/10.1007/978-3-030-14584-2_1
10. Encyclopdia Britannica, available online at https://www.britannica.com. Accessed 1 Feb 2021
11. Erdmann Hilty L, Goodman J, Arnfalk PL (2004) The future impact of ICTs on environmental sustainability. In IPTS Publications

12. Evans S, Vladimirova D, Holgado M, Van Fossen K, Yang M, Silva EA, Barlow CY (2017) Business model innovation for sustainability: towards a unified perspective for creation of sustainable business models. Bus Strateg Environ. https://doi.org/10.1002/bse.1939
13. FAO (2020) The state of world fisheries and aquaculture 2020. Sustainability in action. In European Commission
14. Goralski MA, Tan TK (2020) Artificial intelligence and sustainable development. Int J Manag Educ. https://doi.org/10.1016/j.ijme.2019.100330
15. Hansen EG, Grosse-Dunker F, Reichwald R (2009) Sustainability innovation cube—a framework to evaluate sustainability-oriented innovations. Int J Innov Manag. https://doi.org/10.1142/S1363919609002479
16. Kouhizadeh M, Sarkis J (2018) Blockchain practices, potentials, and perspectives in greening supply chains. Sustainability (Switzerland). https://doi.org/10.3390/su10103652
17. Lele A (2019) Cloud computing. In: Smart innovation, systems and technologies. https://doi.org/10.1007/978-981-13-3384-2_10
18. Pedersen ERG, Gwozdz W, Hvass KK (2018) Exploring the relationship between business model innovation, corporate sustainability, and organisational values within the fashion industry. J Bus Ethics. https://doi.org/10.1007/s10551-016-3044-7
19. Rauter R, Globocnik D, Perl-Vorbach E, Baumgartner RJ (2019) Open innovation and its effects on economic and sustainability innovation performance. J Innov Knowl. https://doi.org/10.1016/j.jik.2018.03.004
20. Rohatgi A, Scherer R, Hatlevik OE (2016) The role of ICT self-efficacy for students' ICT use and their achievement in a computer and information literacy test. Comput Educ. https://doi.org/10.1016/j.compedu.2016.08.001
21. Sharma R, Kamble SS, Gunasekaran A, Kumar V, Kumar A (2020) A systematic literature review on machine learning applications for sustainable agriculture supply chain performance. Comput Oper Res. https://doi.org/10.1016/j.cor.2020.104926
22. Tijan E, Aksentijević S, Ivanić K, Jardas M (2019) Blockchain technology implementation in logistics. Sustainability (Switzerland). https://doi.org/10.3390/su11041185
23. United Nations (2021) Department of Economic and Social Affairs—Sustainable Development, available online at https://sdgs.un.org/goals. Accessed 10 Feb 2021
24. Woschank M, Rauch E, Zsifkovits H (2020) A review of further directions for artificial intelligence, machine learning, and deep learning in smart logistics. Sustainability (Switzerland). https://doi.org/10.3390/su12093760

Implementing Life Cycle Sustainability Assessment in Building and Energy Retrofit Design—An Investigation into Challenges and Opportunities

Hashem Amini Toosi, Monica Lavagna, Fabrizio Leonforte, Claudio Del Pero, and Niccolò Aste

Abstract The built environment is known as a major contributor to both sustainability problems and solutions. Life Cycle Sustainability Assessment (LCSA) which is a promising approach to evaluating the environmental, economic, and social dimensions of building performance, is progressively drawing the building researcher's attention. This chapter aims to review the roots and evolution of building sustainability assessment and discusses the associated challenges of LCSA in building and energy retrofit design. Through a critical review, different assumptions and limitations will be reviewed, and the main challenges of integrating LCSA into building energy retrofit design will be classified and discussed. In the end, the new research lines such as developing integrated LCSA models, application of optimization methods, and Building Information Modeling (BIM) in LCSA will be discussed.

Keywords LCSA · LCA · LCC · SLCA · Decision-Making · Integrated Models · BIM

H. Amini Toosi (✉) · M. Lavagna · F. Leonforte · C. Del Pero · N. Aste
Architecture, Built Environment and Construction Engineering Department, Politecnico Di Milano, Via Ponzio 31, 20133 Milano, Italy
e-mail: hashem.amini@polimi.it

M. Lavagna
e-mail: monica.lavagna@polimi.it

F. Leonforte
e-mail: fabrizio.leonforte@polimi.it

C. Del Pero
e-mail: Claudio.delpero@polimi.it

N. Aste
e-mail: niccolo.aste@polimi.it

S. S. Muthu (ed.), *Life Cycle Sustainability Assessment (LCSA)*,
Environmental Footprints and Eco-design of Products and Processes,
https://doi.org/10.1007/978-981-16-4562-4_6

1 Introduction

The built environment is known as a significant contributor to both sustainability problems and solutions [1]. In such a context, the growing consensus about three facts is of paramount importance leading to progressive efforts in providing comprehensive standards and guidelines for the Life Cycle Sustainability Assessment (LCSA) in the building sector. First is the perception of sustainability as a multidimensional, interdisciplinary, and dynamic science. It requires continuous research to deliver a balanced understanding among various dimensions, including environment, economy, and social dimensions [2, 3]. The dynamism among sustainability pillars and their inherent intricacies demands providing up-to-date standards and guidelines as an indispensable requirement of the assessment works and continuous methodological development and improvement [3–5].

Second is the fact that the expansion of the building sector in response to the growing needs of housing and urbanization trends shows that the building sector is a key role player to achieve the sustainability targets in the present and the future [6, 7]. In the same context, the large share of existing buildings discloses the significant potential of building refurbishment strategies to reach sustainability in this sector [8].

Last but not least is that the building sector can no longer be considered as lineated life products. The building sector's life span is now being studied in the cradle to grave circular scheme by which its sustainability must be evaluated with a whole life cycle perspective [1].

However, various building sustainability assessment frameworks and standards have been released worldwide, a survey on their implementation level in the recent scientific publications is worthy of investigation. This research aims to review and discuss recent scientific publications in LCSA on building energy retrofitting. The goal is to enlighten the extent to which the current standards have been employed in the reviewed publications, the missing aspects (not developed in the standards), and propose scenarios for development and methodological implementation.

To achieve the purpose mentioned above, the published research papers in the field of LCSA in building energy retrofitting are critically reviewed and discussed with a focus on the scope and indicators coverage, the adopted assessment methodologies, weighting, and aggregation methods among LCSA pillars and the final decision-making procedures. The papers are reviewed in four categories containing LCA, LCC, SLCA, and multidimensional LCSA studies where the limitations and advances are critically discussed.

The initial results of the present review highlight that the main challenges in the published research papers could be attributed to (1) lack of required databases, (2) quantification and measurement problems in LCSA impact categories, (3) lack of including impact categories and indicators suggested by standards, and consequently (4) an unbalanced level of development in LCSA pillars evaluation. Also, the level of information provided in each study, both as the input data and the final results, do not match the standards recommendations. Moreover, the development of SLCA assessment methodologies and the synthesis among LCSA pillars through weighting and aggregation are found as vivid obstacles of LCSA application in the literature. These findings along with more issues found in this review, enlighten and explain the critical challenges of LCSA standards implementation in building energy retrofit studies. The data and information management, alongside the considerable computational time required for the in-depth assessment, are the main obstacles in applying LCSA in building energy retrofitting.

In this chapter, the root and evolution of the concept of sustainability over the last decade are studied. Later on, the advent of life cycle sustainability assessment in the building sector and its integration into decision-making methods in building design is reviewed and analyzed. Afterward, The application of LCSA in building energy retrofitting is critically discussed to highlight the main challenges and emerging opportunities in this field of study. The chapter continues with a detailed classification of the main observed challenges of LCSA implementation in building and energy retrofit design. In the end, after an in-depth review, the development of integrated frameworks coupled with optimization models and the integration of Building Information Modeling (BIM) is discussed as promising solutions for implementing LCSA into building and energy retrofit design.

2 Sustainability and Development; The Roots and Evolution

Although the term sustainability is widely used, there are still ambiguities and complexities in the concept of sustainability and sustainable development [9, 10]. This vagueness has been discussed over the years, and it is still being addressed in the academic environment. One possible reason that leads to these complexities is the fact that both sustainability and development are not static and have been evolving in response to the existing dynamism between society and nature [3–5]. The concept of sustainable development is initially driven from economic discipline in [11], where the concerns were about the capacity of limited natural resources to support the increasing human population. According to the Scopus database, the term sustainability was found in 70th, when it first emerged in the scientific literature of economic studies [11], however, some previous studies indicate that the use of this term dates back to a monograph published in 1713 to address the sustainable use of forest resources [2, 12].

As described in dictionaries, development refers to the gradual growth to become more advanced [13]. To clarify this general definition, several theories and interpretations have been proposed by scholars in different fields. One of the definitions collected by [5] elaborates development as a multidimensional process in which major changes in social structures, attitudes, and institutions as well as economic growth, inequality reduction, and poverty eradication are involved. Regarding the historical definitions, the term sustainability primarily addresses the economic-environmental aspects, while development is more oriented to socio-economic issues. Therefore, sustainable development could be interpreted as a concept addressing social, economic, and environmental issues. A similar interpretation is now widely accepted and used.

It is possible to track the efforts to interpret and standardize the term sustainability or sustainable development in the twentieth century. The United Nations conference on the human environment held in Stockholm, Sweden, in 1972, is known as the first international conference to deal with the concept of sustainability [14]. In the declaration of the Stockholm conference, 26 principals were agreed concerning human rights and responsibilities with respect to the social, environmental, and economic aspects. These principles demand an internationally collaborative action plan to achieve sustainable development principles worldwide. In this conference, 109 recommendations were provided to determine how the international participant should effectively regulate their actions to protect the human environment [15].

The World Commission on Environment and Development (WCED) provided the first definition of sustainable development in 1987. In the draft published by WCED, Sustainable development was defined as a *development that meets the needs of the pursuant without compromising the ability of the future generations to meet their needs*. This definition considers the limited ability of the environment to provide the present and future needs of humanity while highlighting that the economic and social requirements in all countries must also be defined in terms of sustainability [16].

An important UN Conference on Environment and Development (UNCED) was held in Rio de Janeiro, Brazil, in 1992 [17]. It is known as the first attempt to implement sustainable development from concept to an international action plan [12]. In the 4th principle of the Rio declaration, environmental protection is emphasized as an integral part of development. The 5th principle refers to eradicating poverty and standard of living as indispensable social requirements of sustainable development. The 12th principle promotes a supportive and open international economic system leading to economic growth for better addressing the problems of environmental degradation [18]. An overall review of the Rio declaration principles shows that environmental issues are the main concerns and the core of this declaration since most of the principles have aimed to promote practical environmental protection actions. Later in 2002, in the World Summit on Sustainable Development held in Johannesburg, South Africa, the balance between economic development, social development, and environmental development as interdependent and mutually reinforcing pillars of sustainability were reaffirmed [19].

Regarding the conflicts among three sustainability pillars and the unequal or insufficient progress in the three dimensions of sustainability, the United Nations Conference on Sustainable Development, Rio + 20 was held in Rio de Janeiro, Brazil, in 2012 [20], emphasizing the balance among sustainability pillars. In the report of this conference entitled "The future we want", poverty eradication is recognized as the greatest global challenge facing the world and an indispensable requirement for sustainable development [21]. In this report, the concept of the green economy is recognized as an essential tool that is available for achieving all pillars of sustainable development. A particular focus on the social aspect highlighted in this conference was also among the Millennium Development Goals, where six goals out of all eight proposed sustainable development goals were oriented to the social dimension of sustainability [22].

The following United Nations Sustainable Development Summit was held in New York in 2015. The resolution adopted by the General Assembly in UN on 25 September 2015 entitled "Transforming our world: the 2030 agenda for sustainable development, changed the traditional concept of sustainable development fundamentally [12] and set out 17 areas of sustainable development goals [23]. These areas are known as the last versions of Sustainable Development Goals (SDGs) declared by United Nations. The year 2015 was a distinguished historical point when the UN set out the 17 SDGs. Not only this step forward for better understanding SDGs and providing the bases of intergovernmental collaboration, but also the Paris Agreement on international effort to increase the abilities of countries in controlling the impacts of climate change [24], have been considered as the historical human efforts to build a more sustainable future.

As elaborated in this section, the definitions of sustainability and development have been subjected to several changes in their meaning and priorities over the last decades. The dynamism and evolving interaction between human society and the natural resources as a complex system could be recognized as the main reason for changing interpretation to define meanings and priorities in sustainability and development.

3 Sustainability Assessment of Buildings—A Life Cycle Approach

The concept of sustainable development targets all human activities and aspects of life and is expected to be adopted by public policy makers to regulate the socio-economic aspect of worldwide activities. It is particularly promoted and applied to address the issues related to the design of the built environment in the last decade [25].

The increasing need for housing in human societies resulting from population growth has led to a rapid expansion of the built environment [26]. The share of the building sector in final energy consumption and GHG emission are increasing worldwide. According to the statistics, the building sector is accounted for 36 and 39% of the final energy consumption and CO_2 emission globally [27]. These values have been estimated at 40 and 32% in European Union (EU), respectively [28].

Given the noticeable contribution of the built environment expansion to the environmental impacts [6, 7], economy, and societies [29] as three pillars of sustainable development [30], growing attention to this issue is now emerging in academies, industries, and policy programs. The increasing awareness about the building sector's considerable impacts on sustainability targets resulted in establishing standards and guidelines to reduce the environmental impacts in this sector. In this context, both the economic and social performance of the building sector has been pursued by emerging studies, as well as the environmental performance, to provide harmony and balance among three life cycle sustainability pillars [1]. However, the social dimension is the least addressed aspect of building sustainability in the literature [31], mainly due to simplifying the sustainability concept in buildings and reducing it to merely environmental sustainability. The terms sustainability and green buildings have been in use interchangeably in building science literature [31]. Consequently, the initial understanding of the term "green" as *"building design strategies that are less environmentally and ecologically damaging than typical practices"* [32], as well as the fact that environmental performance has been better surveyed and standardized [33], could be recognized as the main reasons explaining why all three sustainability dimensions are not equally developed and investigated in the building science literature.

Like the general concept of sustainability, the definition of sustainability in buildings has experienced various interpretations and evolution over the last decade. However, at least three sustainability pillars, such as environment, economy, and society, are now recognized as the widely accepted interpretation; the value judgment among these three pillars is still controversial. Looking at Green Building Rating Systems (GBRSs) such as LEED and BREEAM, the dominancy of a tendency to the environmental interpretation of green or sustainable building is observable [32].

On the other side, several guidelines have been published to standardize the assessment method of the sustainability in buildings with a life cycle approach such as BEES models [34] or the EN standards, including the framework of building sustainability assessment [35], the framework of environmental [36], social [37], and economic performance assessment [38]. However, these methods provided useful methods to measure the building performance regarding the sustainability pillars, but do not address how to make decisions systematically among alternatives with different environmental, economic, and social performances. For instance, BEES models propose a weighted-sum approach to assign a final index to each alternative based on its environmental and economic performance but stay silent about the weighting methods between economy and environment. Likewise, the EN standards have standardized the calculation methods to measure the environmental [39], economic [40], and social performance [41]; they do not clarify how the decision maker should compare different alternatives having conflicting results for each sustainability pillar.

The sustainability pillars in buildings are still being developed and discussed. For instance, looking at GBRSs, there are various aspects and credits, such as the integrative process in LEED or technical quality and process quality in DGNB, that cannot be attributed to the three traditional sustainability pillars. Likewise, recently a fourth dimension has drawn researchers' attention in the literature as institutional dimension [42, 43] that is defined as *"the results of interpersonal processes, such as communication and co-operation, resulting in information and systems of rules governing the interaction of members of a society"* [32].

The various interpretations of sustainability and the lack of accurate definition and calculation methods to quantifiably measure the sustainability pillars show that the sustainability assessment and life cycle sustainability assessment [44] in buildings are still open challenges that need to be more surveyed in the future studies.

4 Life Cycle Sustainability Assessment in the Decision Context—Challenges and Opportunities of Decision-Making Models in Building LCSA

The term Decision-Making (DM) model first emerged in the scientific publication of political science in 1959 [45, 46]. The application of this concept then got increasing attention in other fields as well. As a piece of evidence, the number of scientific publications referred to decision-making models/methods has increased from 2 to more than 15,500 between 1952 to 2020, with a significant growth rate over the recent years. As much as more complicated criteria entered human life, the higher necessity of comprehensive methods to make intelligent decisions is perceived. As a result, the application of DM methods is now widely accepted and is spreading to all fields from very early practice in politics [45] to recent implementation in advanced technologies [47].

Decision-making models are known as the most important application of mathematics in various human activity fields [48]. The necessity of advanced decision-making models arises when at least two assessment criteria exist, and these criteria are contradictory, or the solutions need a value judgment by stakeholders who might have conflicting interests [49].

Facing the global questions that encompass conflicting criteria, multiple diverse goals, contradictory interests, and targets with several different perspectives, Multi-Criteria Decision-Making Models (MCDMs) have been widely implemented to find the appropriate answers to contradictory questions [50]. Sustainability is of those areas that MCDMs models are applied to find comprehensive optimal solutions [51]. As already mentioned, sustainability appeared in the scientific literature in the 70th decade, while the first implementation of a DM model into the sustainability studies dated back to 1997, where it was applied to address the sustainability of future perspective of Swedish urban water systems [52].

Decision-making models in building life cycle sustainability assessment is a very new field of study compared to the comparatively short history of building LCSA. It also shows that building sustainability and life cycle sustainability of buildings were initially developed without taking all the benefits delivered by DM models. However, as LCSA and sustainability assessment include a higher level of complexity and a broader definition over the preceding years, more attention to implementing DM models emerges in the literature.

Amon all MCDM methods, several reviews concluded that AHP is the most popular and applied method in the literature [53, 54]. A study performed by [55] on MCDMs in sustainable energy development issues highlighted that AHP followed by TOPSIS is the most popular multi-criteria decision-making method in the literature.

This fact is also confirmed in our review of few papers published in the field of MCDMs in building life cycle sustainability assessment between 2010 to 2020. Table 1 summarizes the features of recent publications that have applied DM models in building life cycle sustainability starting from 2010 to 2020. In this review, those publications that applied decision-making models in building LCSA were reviewed to highlight the coverage level of LCSA pillars and find the most utilized decision-making methods. As shown in Table 1, most reviewed papers included all three sustainability pillars to evaluate the performance of different types of building design solutions such as structural systems and materials, HVAC systems, and building technologies.

Analytical Hierarchy Process (AHP) is found as the most applied MCDM method within the reviewed papers, while some authors have proposed hybrid DM methods to overcome the drawbacks of the single techniques in their studies [56, 57]. AHP was firstly developed by [58]. According to its developer, AHP is defined as "*a theory of measurement through pairwise comparisons and relies on the judgments of experts to derive priority scales.* Saaty [59] proposed to decompose the decision process into four steps by which it would be possible to apply AHP in making decisions. These steps are [59]: (1) Definition of the problems and determine the kind of knowledge, (2) Structuring the decision hierarchy, (3) Constructing the pairwise comparison matrices, and then (4) Using the obtained weights to define the overall priority. AHP is known as a widely accepted and effective method to support decisions in the complex decision-making process by reducing the problems' complexity through transforming complicated problems into a set of simple comparisons and rankings [57], and increases the transparency and objectivity of decision-making as well as facilitating the detection of controversial items and providing data for establishing agreements [49].

Despite the advantages of the AHP method, one challenging issue associated with this method is that different hierarchies of criteria may affect and cause changes in weight allocations [55, 60]. Cinelli et al. [61] have concluded that although AHP is simple to understand and is well-supported by tools, as a drawback, it is cognitively demanding for decision makers' perspectives.

AHP assumes a full compensation among the criteria that means a low performance in one criterion could be entirely compensated by the high performance of other criteria [61]. While AHP is found as the most applied MCDM method to determine weights of criteria, TOPSIS is known as one of the most popular methods to rank alternatives in a decision-making process, thanks to its straightforward application [55]. Technique for Order of Preference by Similarity to Ideal Solution (TOPSIS) as developed by [67] is based on the concept that the selected alternative must have the shortest distance to the positive ideal solution while keeping the longest distance from the negative ideal solution [57]. Although TOPSIS is highly appreciated due to its easy application in problems with different sizes, some of its disadvantages are also addressed in the literature, such as not considering the correlation of attributes and its difficulty to weight attributes and keep the consistency of judgments [68].

These facts as fundamental critics about the most applied MCDMs in sustainability studies partially show why integrating MCDM methods in this research field could be

Table 1 The features of the recent application of DM models in building LCSA studies

Authors	Goal of analysis	Sustainability Pillars			DM model	Note
		Environment	Economy	Society		
Chandrakumar et al. [56]	Sanitation systems	✓	✓	✓	Fuzzy Analytical Hierarchy Process (FAHP)	Weights based on experts' views
Bakhoum and Brown [57]	Structural materials	✓	✓	✓	Hybrid AHP–TOPSIS–entropy methods	AHP to define weights, TOPSIS to rank alternatives, Entropy to evaluate weight factors of each phase of materials' life cycle
Liu and Qian [62]	Modular construction-based, Semi-prefabricated, Conventional method	✓	✓	✓	AHP-ELCTRE III	Weighting by CFPR-based AHP process, Ranking by ELECTRE method
Rashidi et al.[63]	Construction material supply chain	✓	✓	✓	Multi-attribute decision-making model, TOPSIS	AHP to define weights, TOPSIS to rank alternatives
Arroyo et al.[64]	HVAC systems	✓	✓	✓	Choosing By Advantages (CBA)	Weights assigned according to stakeholders' preferences
Medineckiene et al.[65]	Heating systems of buildings	✓	✓	✗	Complex Proportional Assessment (COPRAS)	AHP to define weights
Wang et al.[66]	Sustainable design options (building tech)	✓	✓	✗	MCDM	Direct weighting via questionnaire

called an open challenge. It is important to note that this paper does not aim to review all MCDM methods; in fact, the pros and cons of the most popular methods have been briefly discussed to understand the most common challenges of implementing decision-making models in sustainability assessment.

5 Implementing Life Cycle Sustainability Assessment in Buildings—The Case of Building Energy Refurbishment

This section reviews the published papers that addressed at least one dimension of life cycle sustainability assessment in building energy retrofitting to understand the methodological advancement, limitations, and challenges in this topic. Therefore, all the relevant papers published and indexed in Scopus and Elsevier until 2020 were retrieved and initially classified into three rubrics, including LCA, LCC, and SLCA studies. These papers are analyzed to clarify the adopted methodologies in each paper to provide a clear picture of the state of the art.

Figure 1 represents the number of papers published between 1989 to 2019 and their distribution around the world. As it is shown in this figure, the publication in this field is increasing fast during preceding years. The European countries, led by Sweden, Italy, Spain, and Portugal, followed by United States, Canada, and China, have the largest share in the research and publication of this field. The lack of LCSA research in several countries underlines that this research field is still not applied worldwide despite its significance in understanding global sustainability issues.

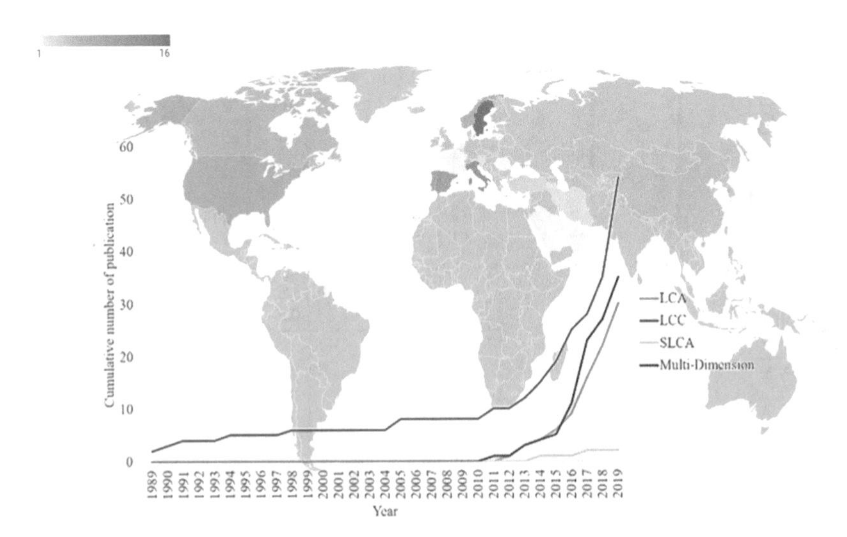

Fig. 1 Cumulative number of publications and their distribution in the world

5.1 LCA in Building Energy Retrofitting; A Review on Methods and Assumptions

Different assumptions and limitations in LCA studies make them difficult to be compared and interpreted against each other. A review is needed to be performed to highlight these assumptions, challenges, and new advancements within the research works related to LCA of building energy retrofit. This review provides essential bases to develop a comprehensive methodological framework for the application of LCA in building energy retrofit design.

Those papers that addressed the LCA of building energy retrofitting in the title or abstract were collected. By reviewing methods and materials in each paper, the challenges of LCA in building energy retrofit are discussed in this section. This part focuses on investigating uncertainties, inconsistencies, challenges, and methodological advances in LCA application in energy retrofit projects. These challenges might affect the reliability and comparability of LCA studies. Reviewing the assumptions and solutions in previous studies will provide a better perspective on how each LCA study could be integrated into the decision-making process of an energy retrofit design project.

One reported issue in previous LCA studies is how to standardize functional unit (FU) in LCA [69] of energy retrofitting which will be reviewed and discussed in this section. The different functional unit has been used in the reviewed papers. In the present review paper, four different kinds of functional units are found:

1. The energy demand/ consumption to provide the required level of thermal comfort [70].
2. The quantity of used materials in a system [71].
3. The unit of area or volume of the refurbished building [72].
4. The whole building under LCA [73].

According to the LCA standards, the functional unit must be clarified in the assessment report. In this review, we realized that some authors have not clearly shown the functional unit in their works, making their results impossible to be compared with other studies [39].

It is reported that the most popular system boundaries in LCA studies are cradle to grave [74]; this statement is also concluded and confirmed in the present review. Some researchers have limited the system boundary of the study solely to the overwhelming life cycle phases [75–77]. For instance, Mangan and Oral [78] limited their analysis to the production stage and use stage due to the lack of data in demolition and end-of-life stages. The system boundary limitation in the research is justified by the fact that previous studies have proven that these eliminated stages (demolition and end of life) have nearly 1 percent of total energy consumption in a building life cycle.

Some other researchers have included the whole building life cycle following the EN 15,978 standard [73, 79–81]. Regarding the reviewed papers in this section, it is found that most researchers have used the whole life cycle phases in their studies, while the lack of databases alongside the negligible impacts are seen as the main

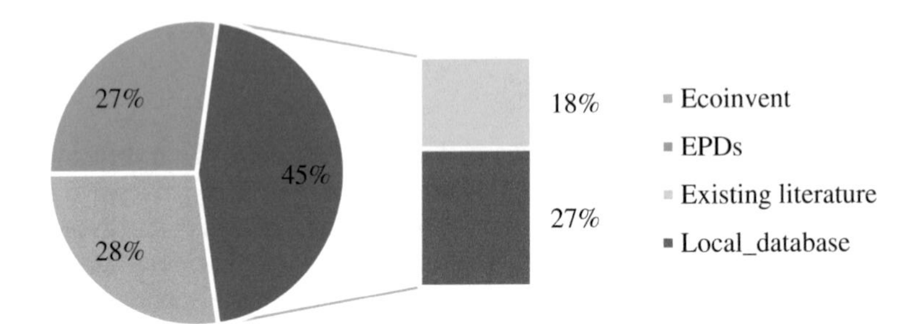

Fig. 2 Life cycle inventory databases used in the reviewed papers

reasons and justification for excluding some life cycle phases in the rest of the reviewed papers. An interesting research by Oregi et al. [82] showed a simplified LCA in which only the production and operation phases are covered could provide accurate results in designing energy retrofit scenarios.

LCI is known as one of the most complicated steps of an LCA study because of the vast numbers of inputs and missing data on materials and building components' environmental performance. In this review, some databases, such as Ecoinvent and EPDs, as well as existing literature or specific data reported by manufacturers, are found as the most common databases (Fig. 2). According to Oregi et al. [80], since different databases may have been prepared using various assumptions, it is essential to pay attention to the possible inconsistency of databases used in research.

Several environmental impact categories and indicators are proposed by LCA standards [39]; however, most published papers have only evaluated a small number of environmental impacts. It is stated that energy and global warming potential (GWP) is the most surveyed key performance indicator in previous studies; however, it is worthy of focusing and reviewing papers that have taken into account more indicators and study how they have been compared against each other. Most of the reviewed papers have only analyzed less than three environmental impact categories mainly due to simplifying the data acquiring procedure. Global warming potential and energy are the most evaluated impacts, as illustrated in Fig. 3.

De Larriva et al. [70] included two environmental indicators, Gross Energy Requirement (GER), and Global Warming Potential (GWP). They have stated that since the LCA is increasingly motivated by the climate change debate, they have chosen these two indicators.

The environmental impact categories in the study performed by Garcia-Perez [71] are limited to global warming potential and embodied energy. Ghose et al. [72] selected twelve environmental impacts recommended by EN 15,978 such as global warming potential, ozone depletion potential, photochemical oxidation potential, acidification potential, eutrophication potential, abiotic depletion (resources and fossil fuels) according to the CML impact assessment method. They also used UseTox method to evaluate human toxicity carcinogenic, human toxicity non-carcinogenic and ecotoxicity freshwater and ILCD 2011 + ReCipe method for particulate matter formation and ionizing radiation. The selection of these categories is in line with

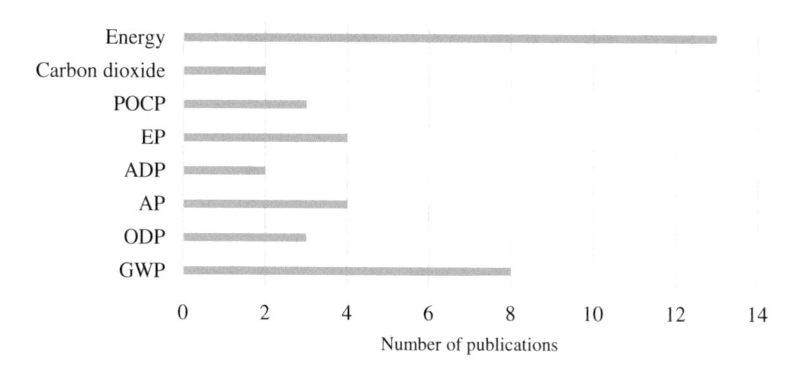

Fig. 3 Number of research papers that addressed each environmental impact. *Note* Energy refers to cumulative energy demand, non-renewable primary energy, embodied energy, life cycle energy and gross energy requirement

national recommendations in New Zealand, as they reported. In contrast, Mangan and Oral [78] and Marique and Rossi [79] only focused on life cycle energy and CO_2 emission. Oregi et al. [80] included only NonRenewable Primary Energy resources in their life cycle impact category. Valancius et al. [77] also included limited environmental indicators such as nonrenewable primary energy and CO_2 emission. Indicators in the study performed by Tadeu et al. [83] are limited to nonrenewable primary energy and greenhouse gas emissions over the building's life cycle.

For simplicity, Oregi et al. [82] considered only one indicator, which is "Use of nonrenewable primary energy sources." Managn and Koclar Oral [76] only took into account LCE and $LCCO_2$ in their study. In the analysis performed by Valacius, Vilutiene, and Rogoza [77], only CO_2 emission and nonrenewable primary energy consumption over the building life cycle are taken into account. The research performed by Nydahl et al. [81] is focused on two environmental impact categories, including life cycle energy use and greenhouse gas emissions. Beccali et al. [73] considered six environmental impact categories at the level of mid-point indicators, including Cumulative Energy Demand (CED), Global Warming Potential (GWP), Ozone Depletion Potential (ODP), Acidification Potential (AP), Eutrophication Potential (EP), Photochemical Ozone Creation Potential (POCP).

The above-mentioned examples also confirm that, although most researchers have followed the LCA standards to calculate environmental impacts, only a few papers have analyzed all proposed environmental impacts by LCA standards. This limitation is mainly due to the lack of databases or with the aim of simplifying the calculation steps, which hopefully will be resolved by developing LCI databases and advancing the LCA software to facilitate the calculation process for non-expert users.

Finally, in some of the reviewed papers, some criteria that are almost neglected in the literature such as the different energy mixes in the future, have also been considered. Ghose et al. [72] have taken into account different energy mixes since, according to national energy programs, the share of fossil fuels is predicted to be reduced by implementing renewable energy sources in New Zealand.

5.2 *LCC in Building Energy Retrofitting; Indicators and Economic Parameters*

This section concentrates on the application of Life Cycle Cost (LCC) as a well-established method for the analysis of the economic performance of buildings [8, 84, 85]. The main parameters of LCC analysis in selected reviewed papers are discussed in this section, including the LCC indicator and economic parameters such as discount rate and energy price inflation rate in each paper.

Several economic indicators such as Net Present Value (NPV), Payback Period, Net Saving or Net Benefit, Saving to Investment Ratio, and Adjusted Internal Rate of Return are proposed by relevant standards [40]. Our review showed that NPV is the most used economic indicator in the reviewed papers (Fig. 4). Other economic indicators such as Value at Risk, Energy productivity, Net Present Cost, Net Saving, Saving to Investment Ratio, Adjusted Internal Rate of Return, Simply Pay Back Period are also adopted in different papers.

Taking accurate Discount Rate (DR) and Inflation Rate (IR) values is of paramount importance in economic assessments. A wide range of values both for the discount rate and inflation rate is found in the reviewed papers, while the EN 16,627:2015 proposes using the discount rate equal to 3 percent for the sake of comparability of the results of LCC studies. Some researchers have compared the LCC results by taking various values for discount and inflation rates in their studies [86–90]. For instance, Copiello, Gabrielli, and Boniaci [91] reported that the discount rate might affect the results four times as much as the energy price. They also mentioned that the discount rate might also affect the energy retrofit project by encouraging owners for higher initial investment. Our analysis shows that the values taken by researchers are usually higher than the 10-year average values which are reported by the countries.

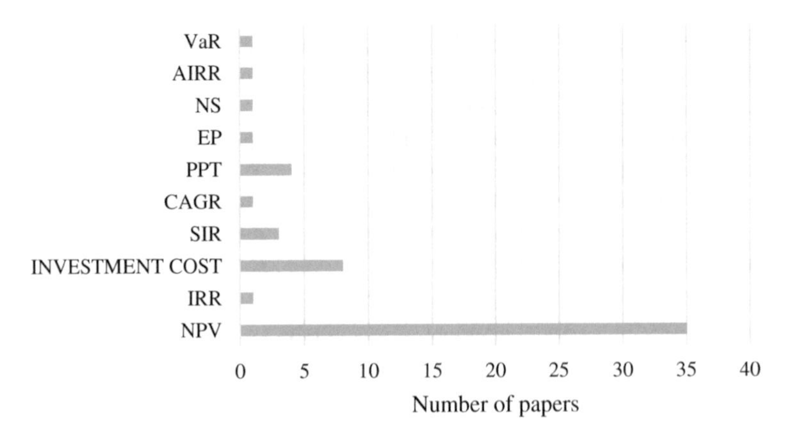

Fig. 4 Number of research papers that addressed each LCC indicator, Var: Value at Risk, AIRR: Adjusted Internal Rate of Return, NS: Net Saving, EP: Energy Productivity, PPT: Payback Period Time, CAGR: Compound Annual Growth Rate, SIR: Saving to Investment Ratio, IRR: Internal Rate of Return, NPV: Net Present Value

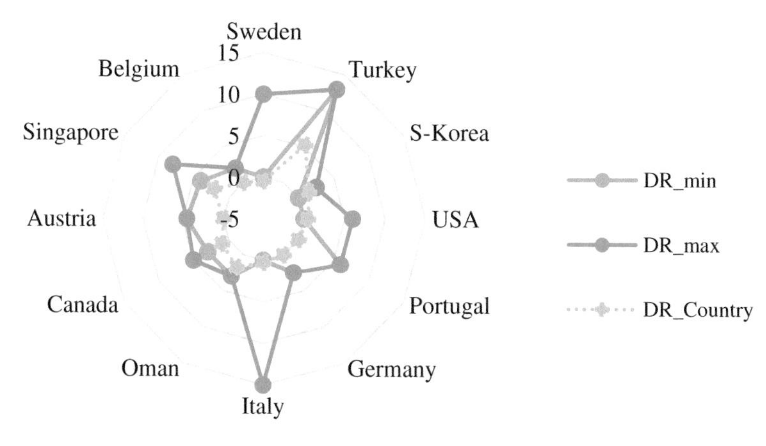

Fig. 5 The minimum, maximum discount rate (%) applied in the LCC studies in each country versus the 10-year average discount rate of the countries

As shown in Fig. 5, the minimum and maximum values of the discount rate applied in the research papers are higher than the actual average value of the discount rate in each country. Although it is worthy of investigation to analyze the influence of various DR values in research works, it is recommended to adopt the macroeconomic values according to the actual economic situation of the study project. Moreover, in compliance with the EN standards taking similar DR values in LCC studies in the building sector increases the comparability of the results. In case the researchers aim to conduct sensitivity analysis to evaluate the impact of the different economic situations on their project, the economic parameters should also represent the actual values in the projects' economic contexts and the relevant standards (e.g., Italian studies in Table 2, Figs. 5 and 6).

Regarding the values of discount rate and energy price inflation rate, Fig. 7 represents important information about the reviewed papers. As it is illustrated, most research works are performed with an energy price inflation rate lower than the average in all papers. However, a variety of discount rate values are considered in papers. Figure 7 shows that the papers published in different countries tend to conduct LCC analysis with a combination of low to a medium value of inflation rate and medium to high value for discount rate.

5.3 SLCA in Building Energy Retrofitting; The Implementation Level

Any adverse or beneficial change to the society or the quality of life that could be expressed with quantifiable indicators is defined as SLCA impact with respect to the following categories in EN 15,643-3:2012 [37]: accessibility, adaptability, health

Table 2 Summarizes the economic parameters applied in selected LCC studies in each country.

Countries	Discount rate			Energy price inflation rate			Country (electricity)
	Min	Max	Country	Min	Max	Country (gas)	
Austria [92]	4.5	4.5	0	–	–	–	–
Belgium [93]	2	2	0	3	3	4.392	4.39
Canada [94]	3	5	1	–	–	–	–
China [84, 95]	6.6	8					
Germany [96]	2.5	2.5	0	0	4	0.955	2.86
Italy [87, 88, 97–99]	0	15	0.25	0	4.5	0.562	2.104
Oman [100]	3	3	1.726	–	–	–	–
Portugal [101]	6	6	0	4	4	3.238	3.329
Singapore [89]	4	8	2.15				
Sweden [102–109]	0	10	−0.5	0.5	3	1.33	0.759
S. Korea [8, 110, 111]	0	2.54	1.5	–	–	–	–
Turkey [112–114]	13	13	5.25	–	–	–	–
United States [86, 115, 116]	0	6	0.5	5	5	2.233	0.6

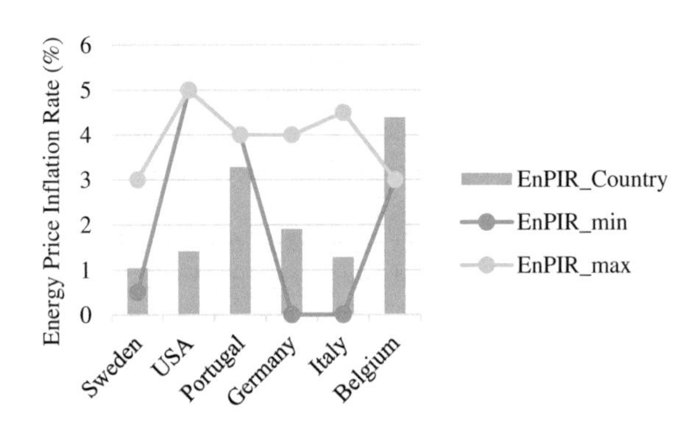

Fig. 6 The minimum, maximum Energy Price Inflation rate (%) applied in the LCC studies in each country versus the 10-year En-PIR of the countries (the average of gas and electricity price for end-user in residential buildings)

and comfort, loading and neighborhood, maintenance, safety/security, sourcing of material and services, and stakeholder involvement.

In the present review, no published research paper is found which directly addresses the social dimension of building energy retrofitting with a life cycle approach in the title, abstract, or keywords. However, few papers are found in which some social indicators such as thermal comfort [117], human live risk [118], and social feasibility [119] are taken into account. Thermal comfort could be considered as a social aspect of building sustainability assessment according to EN 15,643-3:2012 since it affects occupants and users' satisfaction levels. The studies performed

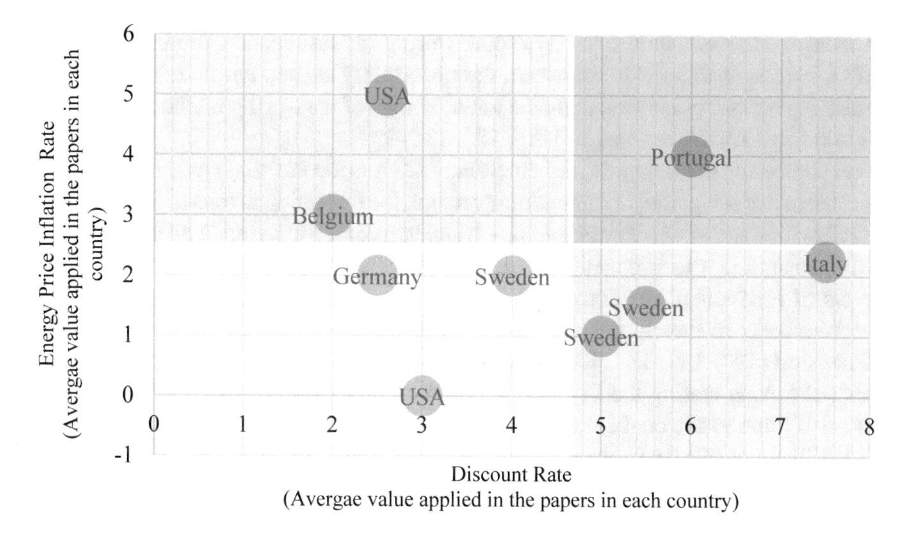

Fig. 7 The average of discount rate versus the average of energy-price inflation rate adopted in selected LCC studies in each country

by Assiego de Lavaria et al. [120] and Mostavi et al. [117] are of those few ones that have addressed one of the impact categories of social LCA in their assessment.

6 LCSA Challenges; A Classification of the Open Challenges and a Discussion on Emerging Solutions in the Literature

Several challenges are associated with sustainability assessment as a multidimensional interdisciplinary field of study [121]. However, taking a life cycle approach to sustainability assessment increases the study's comprehensiveness; it might result in a higher level of sophistication since more databases and assessment methods with a higher level of uncertainties and inconsistencies might be included in the analysis. Given the discussions in the previous sections and the research papers that addressed the sustainability assessment challenges within the last five years, the main challenges of measurement in life cycle sustainability assessment are presented and discussed in this section. Such a discussion helps to enlighten what aspects of LCSA need to be investigated and developed by further research works in the future.

The challenges associated with data collection and accessibility are constantly reported as one of the main obstacles in implementing life cycle sustainability assessment [1]. The required databases to conduct life cycle sustainability assessment are not readily available for specific materials, products, or services around the world [122–129] and the data acquisition procedure is not straightforward [126–134] due to the complexity of the data preparation and data-sharing challenges [124, 135].

Although databases have been developed during the last decade, the lack of data is still a barrier in this field. Moreover, the uncertainty caused by the missing data of emerging technologies alongside the uncertainties of measuring methods are known as important LCSA challenges [126, 134, 136–139].

A significant challenge of implementing LCSA is the fact that no consensus exists to establish or adopt a clear methodology to link the three dimensions of sustainability [1, 122, 129, 140–143]. The combination and harmonization among different metrics and measurement techniques [122, 139, 140, 143] alongside the different maturity levels of assessment methods for LCSA pillars [131, 136, 137, 144], specifically the weakness in developing the quantifiable measurement methods of the social dimension [127, 129, 131, 133, 134, 144–146] is of the most critical challenges in this field. Aggregating the LCSA pillars is a complex issue [1, 122, 126, 129, 136, 140–142] due to the challenges associated with selecting the suitable indicators [3, 122, 125, 127–129, 145], weighting [1, 44, 126, 134, 139, 146–148], normalization [44, 134, 142, 146, 148], and formulating life cycle sustainability [1, 129, 130, 133, 139, 149, 150].

As illustrated in Fig. 8, the associated challenges of measuring LCSA could be initially classified into six groups, including Data, Measuring methods, Aggregation, Indicator selection, Uncertainties, and Results. Further and future research works need to be conducted to resolve these challenges. Apart from the continuous efforts to standardize the measuring methods, to develop databases and reduce the uncertainty of the evaluations through methodological advancements, new trends in the literature are found to answer the challenges of integrating and facilitating LCSA

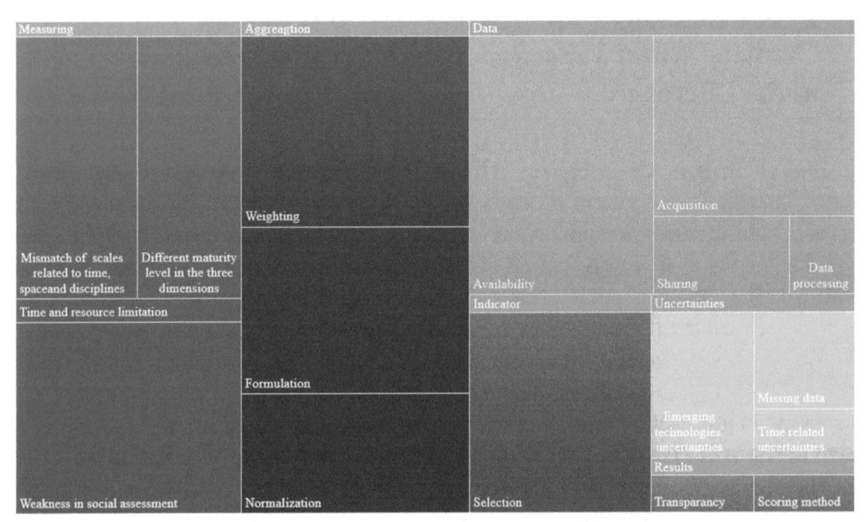

Fig. 8 Classification of the existing challenges related to measuring life cycle sustainability assessment. The size of each section corresponds with the number of papers addressed each challenge

in the building design process. The advent of developing integrated LCSA models and digitalization-LCSA nexus in the literature are examples of the new research trends aiming at providing solutions to ease the LCSA implementation in building and energy retrofitting design.

6.1 Integrated LCSA Models—Multi-dimensional LCSA Studies and Application of Optimization Methods

The research works that have addressed more than one LCSA dimension have constantly been increasing over the last few years. Several authors have included LCA and LCC simultaneously in their analysis, such as Krarti and Dubey [151] evaluated the economic and environmental benefits of three levels of energy retrofitting for different building types, including residential, commercial, and governmental buildings. Ruparathna et al. [152] proposed a method to find the best energy retrofit scenarios of buildings by considering energy consumption, life cycle costs, and GHG emission. Some researchers proposed a conceptual framework for an integrated LCSA model using a weighted-sum approach that includes all three LCSA pillars [153, 154]. Implementing LCSA into energy building energy retrofitting is a multi-objective task for which optimization methods and algorithms to find the extremum values of multi-variable functions are used by several authors over the last years to resolve the complexity of this task [1, 155].

Table 3 summarizes the recent research papers that addressed more than one LCSA pillar and represents the indicators and optimization algorithms adopted in each study.

Although several research papers have already been published addressing multi-dimensional life cycle sustainability assessment of buildings, our review showed that there are still challenges to be resolved. For instance, the lack of well-established quantification methods to measure SLCA is still a barrier to implement LCSA. Moreover, the lack of consensus on weighting methods for aggregating LCSA pillars is still an open challenge in this field. These challenges are expected to be resolved through future research on integration methods; however, it requires the development of LCI databases, measurement development, and standardization of LCSA pillars.

6.2 BIM-Based LCSA—A Solution for Data Management and Processing

Digitalization in the built environment and the application Building Information Modeling (BIM) are growing rapidly in the construction industry and can help

Table 3 Summary of integrated multidimensional LCSA studies

Author	LCA	LCC	SLCA	Indicators	Optimization algorithm
Chantrelle et al. [156]	✓	✓	✓	GC, Energy, CO_2, Thermal comfort	Genetic Algorithm: NSGA-II
Kusar et al. [118]	✗	✓	✓	NPV, Human live risk, Structural safety	✗
Risholt et al. [157]	✓	✓	✓	GC, CO_2,Thermal/ air quality	✗
Gustafsson et al. [158]	✓	✓	✗	GC, PEC, NRE, CO_2	✗
Holopainen et al. [119]	✓	✓	✓	GC, GWP, Social feasibility	✗
Pal et al. [159]	✓	✓	✗	Life cycle carbon footprint and life cycle cost	Genetic Algorithm: NSGA-II
Ramin et al. [160]	✓	✓	✓	Energy, CO_2, cost, water	Multi-objective optimization
Moschetti and Brattebø [161]	✓	✓	✗	NPV, CED, GWP	–
Ylmén et al. [162]	✓	✓	✗	Global warming potential, life cycle costs	Genetic Algorithm
M. Gustafsson et al. [163]	✓	✓	✗	NPV, GWP, Freshwater EP, particulate matter formation, NRPE	✗
Mauro et al. [164]	✗	✓	✓	LCC, Thermal comfort	NSGA-II algorithm
Mostavi et al. [117]	✓	✓	✓	LCE, LCC and Thermal comfort index	HS Algorithm
Almeida and Ferreira [165]	✓	✓	✗	GC, CO_2, PE	✗
Almeida [166]	✓	✓	✗	GC, GWP, NRPE, TPE	
Jokisalo et al. [167]	✓	✓	✗	LCC and Energy consumption	NSGA-II algorithm
Amirhosain and Hamma [168]	✓	✓	✗	LCC, energy consumption	NSGA-II algorithm, ANN, ML
Hirvonen et al. [169]	✓	✓	✗	LCC, CO_2 emission	NSGA-II algorithm
Conci et al. [170]	✓	✓	✗	NPV, GWP	✗
Amini Toosi and Lavagna [154]	✓	✓	✓	NPV, several environmental impacts, Thermal comfort	Genetic Algorithm

(continued)

Table 3 (continued)

Author	LCA	LCC	SLCA	Indicators	Optimization algorithm
Mateus et al. [171]	✓	✓	✗	NPV, GWP, CED	✗

and support the integrated design process through improving information management and cooperation between designers, producers, and end-users during the whole buildings' life cycle stages [172–175].

Buildings consist of various components; this brings a massive amount of information and complexity to the design phase [176]. This is usually reported as the main reason for performing LCSA at the later project phases, where the complexity and uncertainties are reduced [177, 178]. BIM tools are capable of providing and present both graphical, numerical, and descriptive information of buildings in different levels of development (LOD) [179–182], which is an important requirement for applying an LCA during the design phase. It is also reported that the use of LCA methods in the building sector cannot be developed without developing the level of information in this sector, on the other hand, it is stated that the use of BIM for public buildings will be compulsory in the EU from October 2018 [182] and expected to be extensively used in the near future [173]. All these facts indicate that BIM-based life cycle sustainability assessment is a promising and indispensable solution to resolving data integration and management challenges.

Several researchers have addressed the application of BIM in building life cycle assessment. For instance, Malmqvist et al. [183] proposed BIM tools to overcome data analysis problems during the early stages of the design process. They indicated that in the early stages of design, there are many possible solutions and decisions to take, while the precise data which are required for the LCA calculations are usually available at the later design stages. To overcome this problem, speed up the LCA calculation process, and increase the accuracy and completeness of the evaluations, they suggested using BIM tools in the LCA-design process [183].

Many researchers have elaborated the necessity of BIM application in LCA and have tried to use BIM tools in an LCA process [181, 184–187]. Although many studies demonstrate the advantages and benefits of BIM-based life cycle assessment and the integration of BIM and LCA [187–189], serious challenges such as software integration or data requirements are still the main problems and barriers in this field [182]. The existing BIM tools are not capable of comparing different alternatives, also still suffer from data library limitations [74, 184, 186]. It is understandable that in order to make the BIM-LCA integration useful, the input data, assessment process, result acquisition, and interpretation must be as easily achievable as soon as possible, and the whole integrated assessment system must be user-friendly [182].

The integration between BIM tools and energy simulation software is not still fully developed. Also, data exchange between BIM and LCA tools is another critical issue. In some research works, automated produced bills of materials are imported into the excel sheets for the LCA calculation. Ajayi et al. [187], Basbagill et al.

[190], Peng [184], and Houlihan et al. [191], as well as many other researchers, have used a manual process for data exchange between BIM tools and LCA tools or LCA calculation sheet in Excel. However, some plugins on Revit Autodesk make it possible to quantify environmental impacts in the BIM environment based on LCA methods [183, 193]. There is still a gap in software integration between BIM, LCSA, and energy simulation tools.

Another problem stated and confirmed by researchers is that BIM databases are not developed enough for the LCA process. Because of this problem, in most cases, the bill of material quantities and material properties are edited manually by the end-users [182].

Although the BIM concept is not effective in integrating building performance assessment into the sketch design phase due to the excessive required time for modeling [192], if the task is about implementing the LCSA in designing energy retrofit scenarios, BIM can significantly facilitate the assessment process, since many of design parameters have already been defined and the uncertainty is lower in energy retrofit design compared to the sketch and initial design phase.

To start an integrated BIM-LCSA design process, some questions must be answered first:

1. What are the design and assessment goals? The answer will determine what kinds of performance criteria must be assessed.
2. What is the assessment methodology, and what kind of assessment methodologies need to be integrated into BIM?
3. What kind of data and databases need to be integrated into BIM?
4. What is the required detail, accuracy, and completeness level for the performance assessment?
5. How should the result be reported, and in which way should they be processed and used?

As an example to answer one of these questions, Dupuis et al. [193] indicated that to be able to perform an accurate LCA study and achieve sound results, every data element should be at least at the LOD 350 detail level. Each BIM model at a lower level than LOD 350 means that some essential data for LCA calculation will be missed.

Although BIM-based LCSA is a promising solution to overcome the integration challenges of LCSA in building and energy retrofit design, there are still some challenges in applying this framework, such as availability of life cycle inventory (LCI) databases, software integration, and transferring building information between modeling software and LCSA tools. Given the rapid progress in developing design-assessment tools and LCI databases, it is expected that LCSA analyses would be possible to be performed in the BIM environment without using intermediary tools in the near future.

7 Conclusion

This chapter discussed how sustainability assessment has been developed from a single-dimension and environmental-oriented interpretation to a multidimensional interdisciplinary research field by reviewing its roots and evolution path over the last decades. Then we discussed how decision-making models have been integrated into life cycle sustainability assessment of buildings to facilitate the informed decision-making in the multi-objective design-assessment contexts such as building life cycle assessment. Through literature review on the implementation of LCSA in building energy retrofitting, we discussed the existing challenges of the life cycle sustainability assessment. We concluded that different assumptions such as various functional units, system boundaries, and lack of a wide range of standard environmental impacts result in complexities in the comparability of the reviewed research works. Moreover, the lack of LCI databases is known as the main obstacle of LCA application. Also, we showed that the level of documentation of some research works is lower than the recommendation by relevant standards, which need to be considered in future research works to enhance the readability and comparability of the results.

Regarding the LCC studies, we showed that Net Present Value is the most popular economic indicator used by several researchers to evaluate the economic performance of their retrofit design. Macroeconomic parameters such as discount rate and energy price inflation rate adopted in each paper were discussed, and we showed that the assumed values in the research papers are lower than the actual values in the economic context of the study in most cases. However, we highlighted that several papers have taken various values of discount rate and energy price inflation rate to evaluate the impacts of these parameters on the final results.

In this review, the lack of integrating social life cycle assessment into evaluating energy retrofit design is found as one of the main limitations. The SLCA is less developed than LCA and LCC and requires more methodological advancements, especially in developing measurement methods, quantifiable indicators, and databases.

It is also found that the number of multidimensional LCSA studies in building and energy retrofit design is increasing over the preceding years, and several researchers have proposed integrated frameworks to implement LCSA into the building design. In this context, the development of optimization algorithms and available tools are promising solutions to facilitate the LCSA implementation in the multi-objective building design process. Likewise, the BIM-based approach to integrate LCSA into the building design process attracted the researcher's attention for solving the complexity of data management and processing in building life cycle sustainability assessment. Nevertheless, it is essential to develop measurement methods, standardization, and aggregation methods of LCSA pillars alongside providing more comprehensive databases and developing integrated software and tools by future research works to facilitate the implementation of LCSA in the building and energy retrofit design process.

References

1. Amini Toosi H, Lavagna M, Leonforte F, Del Pero C, Aste N (2020) Life cycle sustainability assessment in building energy retrofitting—a review. Sustain Cities Soc 60:102248, Sep. 01, 2020. Elsevier Ltd. https://doi.org/10.1016/j.scs.2020.102248
2. Du Pisani JA (2006) Sustainable development—historical roots of the concept. Environ Sci 3(2):83–96. https://doi.org/10.1080/15693430600688831
3. Komeily A, Srinivasan RS (2015) A need for balanced approach to neighborhood sustainability assessments: a critical review and analysis. Sustain Cities Soc 18:32–43. https://doi.org/10.1016/j.scs.2015.05.004
4. Clark WC, Dickson NM (2003) Sustainability science: the emerging research program. Proc Natl Acad Sci 100(14):8059–8061. https://doi.org/10.1073/pnas.1231333100
5. Mensah J (2019) Sustainable development: Meaning, history, principles, pillars, and implications for human action: Literature review. Cogent Soc Sci 5(1). https://doi.org/10.1080/23311886.2019.1653531
6. Aste N, Caputo P, Buzzetti M, Fattore M (2016) Energy efficiency in buildings: what drives the investments? the case of lombardy region. Sustain Cities Soc 20:27–37. https://doi.org/10.1016/J.SCS.2015.09.003
7. Mirabella N et al (2018) Strategies to Improve the energy performance of buildings: a review of their life cycle impact. Buildings 8(8):105. https://doi.org/10.3390/buildings8080105
8. Song K, Ahn Y, Ahn J, Kwon N (2019) Development of an energy saving strategy model for retrofitting existing buildings: a Korean case study. Energies 12(9). https://doi.org/10.3390/en12091626
9. Mebratu D (1998) Sustainability and sustainable development: historical and conceptual review. Environ Impact Assess Rev 18(6):493–520. https://doi.org/10.1016/S0195-9255(98)00019-5
10. Roostaie S, Nawari N, Kibert CJ (2019) Sustainability and resilience : a review of definitions, relationships, and their integration into a combined building assessment framework. Build Environ 154(March):132–144. https://doi.org/10.1016/j.buildenv.2019.02.042
11. Hartwick JM (1974) Price sustainability of location assignments. J Urban Econ 1(2):147–160. https://doi.org/10.1016/0094-1190(74)90014-X
12. Shi L, Han L, Yang F, Gao L (2019) The evolution of sustainable development theory: types, goals, and research prospects. Sustain 11(24):1–16. https://doi.org/10.3390/su11247158
13. Development (2020) "Development," Oxford online dictionary. https://www.oxfordlearnersdictionaries.com/definition/english/development
14. UN (1972) United Nations conference on the human environment (Stockholm Conference). United Nation. https://sustainabledevelopment.un.org/milestones/humanenvironment
15. United_Nations (1972) Report of the United Nations Conference on the Human Environment. New York
16. Our_Common_Future (1987) Report of the world commission on environment and development: our common future. New York
17. UN (1992) United Nations Conference on Environment and Development (UNCED), Earth Summit (Rio de Janeiro, Brazil). https://sustainabledevelopment.un.org/milestones/unced
18. Rio_Conference (1992) Report of the united nations conference on environment and development. New York
19. Johannesburg_Conference (2002). Report of the World Summit on Sustainable Development (WSSD), Johannesburg Summit. New York
20. UN (2012) United Nations Conference on Sustainable Development, Rio+20. https://sustainabledevelopment.un.org/rio20
21. Rio+20 (2012) Resolution adopted by the General Assembly on 27 July 2012, (The future we want). New York
22. UN (2013) Millennium Development Goals (MDGs). https://www.un.org/millenniumgoals/bkgd.shtml

23. UNSDS (2015) Resolution adopted by the General Assembly on 25 September 2015, Transforming our world: the 2030 Agenda for Sustainable Development. New York
24. UN (2015) The Paris Agreement. https://unfccc.int/process-and-meetings/the-paris-agreement/the-paris-agreement
25. Mattoni B, Guattari C, Evangelisti L, Bisegna F, Gori P, Asdrubali F (2018) Critical review and methodological approach to evaluate the differences among international green building rating tools. Renew Sustain Energy Rev 82(October 2017):950–960. https://doi.org/10.1016/j.rser.2017.09.105
26. Lim YS et al (2015) Education for sustainability in construction management curricula. Int J Constr Manag 15(4):321–331. https://doi.org/10.1080/15623599.2015.1066569
27. UNEP (2018) Global alliance for buildings and construction, 2018 global status report, p 325. https://doi.org/10.1038/s41370-017-0014-9
28. Mateus R, Silva SM, De Almeida MG (2019) Environmental and cost life cycle analysis of the impact of using solar systems in energy renovation of Southern European single-family buildings. Renew Energy 137:82–92. https://doi.org/10.1016/j.renene.2018.04.036
29. Janjua SY, Sarker PK, Biswas WK (2020) Development of triple bottom line indicators for life cycle sustainability assessment of residential bulidings, 264(November 2019)
30. United Nations Environmental Program (UNEP) (2011) Note 12: Towards a L ife C ycle S ustainability A ssessment
31. Kamali M, Hewage K, Milani AS (2018) Life cycle sustainability performance assessment framework for residential modular buildings: aggregated sustainability indices. Build Environ 138(March):21–41. https://doi.org/10.1016/j.buildenv.2018.04.019
32. Doan DT, Ghaffarianhoseini A, Naismith N, Zhang T, Ghaffarianhoseini A, Tookey J (2017) A critical comparison of green building rating systems. Build Environ 123:243–260. https://doi.org/10.1016/j.buildenv.2017.07.007
33. Visentin C, William A, Braun AB (2020) Life cycle sustainability assessment : a systematic literature review through the application perspective , indicators, and methodologies 270. https://doi.org/10.1016/j.jclepro.2020.122509
34. Lippiatt B (2007) BEES 4.0: Building for environmental and economic sustainability, technical manual and user guide, director, p. 307, 2007, doi: 860108
35. CEN/TC-350 CEN/TC-350 (European Committee for Standardization (CEN)) (2010) EN 15643-1:2010, Sustainability of construction works—sustainability assessment of buildings—Part 1: General framework
36. CEN/TC-350 CEN/TC-350 (European Committee for Standardization (CEN)) (2011) EN 15643-2:2011 Sustainability of construction works—assessment of buildings—Part 2: Framework for the assessment of environmental performance
37. CEN/TC-350 CEN/TC-350 (European Committee for Standardization (CEN)) (2012) EN 15643-3:2012 Sustainability of construction works—assessment of buildings—Part 3: Framework for the assessment of social performance
38. CEN/TC-350 (European Committee for Standardization (CEN)), "EN 15643-4:2012 Sustainability of construction works - Assessment of buildings - Part 4: Framework for the assessment of economic performance," 2012.
39. EN_15978 (2011) EN 15978:2011, Sustainability of construction works. Assessment of environmental performance of buildings. Calculation method
40. EN_16627 (2015) EN 16627:2015 Sustainability of construction works. Assessment of economic performance of buildings. Calculation methods
41. EN_16309:2014+A1:2014 (2014) EN 16309:2014+A1:2014, Sustainability of construction works. Assessment of social performance of buildings. Calc Methodol
42. Valentin A, Spangenberg JH (2000) A guide to community sustainability indicators 20:381–392
43. Shari A, Murayama A (2014) Neighborhood sustainability assessment in action: Cross-evaluation of three assessment systems and their cases from the US , the UK, and Japan," vol 72. https://doi.org/10.1016/j.buildenv.2013.11.006

44. Sala S, Farioli F, Zamagni A (2013) Life cycle sustainability assessment in the context of sustainability science progress (part 2), pp 1686–1697. https://doi.org/10.1007/s11367-012-0509-5
45. Frankel J (1959) Towards a decision-making model in foreign polICY. Polit Stud 7(1):1–11. https://doi.org/10.1111/j.1467-9248.1959.tb00888.x
46. Reynolds PA (1959) A comment on Mr. Frankel's article towards a decision-making model in foreign policy. Polit Stud 7(3):302–304
47. Gong B, Liu R, Zhang X (2020) Market acceptability assessment of electric vehicles based on an improved stochastic multicriteria acceptability analysis-evidential reasoning approach. J Clean Prod 269:121990.https://doi.org/10.1016/j.jclepro.2020.121990
48. Medineckiene M, Zavadskas EK, Björk F, Turskis Z (2015) Multi-criteria decision-making system for sustainable building assessment/certification. Arch Civ Mech Eng 15(1):11–18. https://doi.org/10.1016/j.acme.2014.09.001
49. VillarinhoRosa L, Haddad AN (2013) Building Sustainability Assessment throughout Multicriteria Decision Making. J Constr Eng 2013:1–9. https://doi.org/10.1155/2013/578671
50. Karjalainen TP et al (2013) A decision analysis framework for stakeholder involvement and learning in groundwater management. Hydrol Earth Syst Sci 17(12):5141–5153. https://doi.org/10.5194/hess-17-5141-2013
51. Hannouf M, Assefa G (2018) A life cycle sustainability assessment-based decision-analysis framework. Sustain 10(11). https://doi.org/10.3390/su10113863
52. Aspegren G, Hellström H, Olsson B (1997) The urban water system—a future Swedish perspective. Water Sci Technol 35(9):33–43.https://doi.org/10.1016/S0273-1223(97)00182-0
53. Diaz-balteiro L, González-pachón J, Romero C (2017) Measuring systems sustainability with multi-criteria methods: a critical review. Eur J Oper Res 258(2):607–616.https://doi.org/10.1016/j.ejor.2016.08.075
54. Zarte M, Pechmann A, Nunes IL (2019) Decision support systems for sustainable manufacturing surrounding the product and production life cycle e A literature review. J Clean Prod 219:336–349. https://doi.org/10.1016/j.jclepro.2019.02.092
55. Siksnelyte I, Zavadskas EK, Streimikiene D, Sharma D (2018) An overview of multi-criteria decision-making methods in dealing with sustainable energy development issues. Energies 11(10). https://doi.org/10.3390/en11102754
56. Chandrakumar C, Kulatunga A, Mathavan S (2018) A multi-criteria decision-making model to evaluate sustainable product designs based on the principles of Design for Sustainability and Fuzzy Analytic Hierarchy Process A multi-criteria decision-making model to evaluate sustainable product designs based o. Sustain Des Manuf 2017. SDM 2017. Smart Innov. Syst. Technol. Springer, Cham, vol 68, no June, pp 347–354, 2018, doi: https://doi.org/10.1007/978-3-319-57078-5_34
57. Bakhoum ES, Brown DC (2013) A hybrid approach using AHP-TOPSIS—entropy methods for sustainable ranking of structural materials. Int J Sustain Eng Taylor Fr 6(3):212–224. https://doi.org/10.1080/19397038.2012.719553
58. Saaty TL (1980) The analytical hierarchical process. Wiley, New York
59. Saaty TL (2008) Decision making with the analytic hierarchy process, 1(1)
60. Walling E (2020) Developing successful environmental decision support systems: challenges and best practices 264 no(April, 2020). https://doi.org/10.1016/j.jenvman.2020.110513.
61. Cinelli M, Coles SR, Kirwan K (2014) Analysis of the potentials of multi criteria decision analysis methods to conduct sustainability assessment. Ecol Indic 46:138–148. https://doi.org/10.1016/j.ecolind.2014.06.011
62. Liu S, Qian S (2019) Towards sustainability—oriented decision making : Model development and its validation via a comparative case study on building construction methods, no. June 2018, pp 860–872, 2019. https://doi.org/10.1002/sd.1946
63. Alireza Ahmadian FF, Rashidi TH, Akbarnezhad A, Waller ST (2015) BIM-enabled sustainability assessment of material supply decisions. doi: https://doi.org/10.1108/ECAM-12-2015-0193

64. Arroyo P, Tommelein ID, Ballard G, Rumsey P (2015) Choosing by advantages: a case study for selecting an HVAC system for a net zero energy museum Choosing by advantages: a case study for selecting an HVAC system for a net zero energy museum. Energy Build 111(October):26–36. https://doi.org/10.1016/j.enbuild.2015.10.023
65. Medineckiene M, Monitoring HE, Centre A, Turskis Z, Zavadskas EK (2011) Life-cycle analysis of a sustainable building , aplying multi-criteria decision making method, no. October 2014
66. Wang N, Chang Y, Nunn C (2010) Lifecycle assessment for sustainable design options of a commercial building in Shanghai. Build Environ 45(6):1415–1421. https://doi.org/10.1016/j.buildenv.2009.12.004
67. Hwang K, Yoon CL (1981) Multiple attribute decisionmaking: methods and applications. Springer, New York
68. Velasquez M, Hester P (2013) An analysis of multi-criteria decision making methods. Int J Oper Res 10(2):56–66
69. Anand CK, Amor B (2017) Recent developments, future challenges and new research directions in LCA of buildings: a critical review. Renew Sustain Energy Rev 67:408–416. https://doi.org/10.1016/j.rser.2016.09.058
70. Assiego de Larriva R, Calleja Rodríguez G, Cejudo López JM, Raugei M, Fullana i Palmer P (2014) A decision-making LCA for energy refurbishment of buildings: Conditions of comfort. Energy Build 70:333–342. https://doi.org/10.1016/J.ENBUILD.2013.11.049
71. García-Pérez S, Sierra-Pérez J, Boschmonart-Rives J (2018) Environmental assessment at the urban level combining LCA-GIS methodologies: a case study of energy retrofits in the Barcelona metropolitan area. Build Environ 134:191–204. https://doi.org/10.1016/J.BUILDENV.2018.01.041
72. Ghose A, McLaren SJ, Dowdell D, Phipps R (2017) Environmental assessment of deep energy refurbishment for energy efficiency-case study of an office building in New Zealand. Build Environ, 117:274–287, May 2017, Accessed: Dec. 05, 2018. https://www.sciencedirect.com/science/article/pii/S0360132317301105
73. Beccali M, Cellura M, Fontana M, Longo S, Mistretta M (2013) Energy retrofit of a single-family house: Life cycle net energy saving and environmental benefits. Renew Sustain Energy Rev 27:283–293. https://doi.org/10.1016/j.rser.2013.05.040
74. Anand CK, Amor B (2017) Recent developments, future challenges and new research directions in LCA of buildings: a critical review. Renew Sustain Energy Rev 67:408–416. https://doi.org/10.1016/j.rser.2016.09.058
75. Bin G, Parker P (2012) Measuring buildings for sustainability: comparing the initial and retrofit ecological footprint of a century home—the REEP House. Appl Energy 93:24–32. https://doi.org/10.1016/j.apenergy.2011.05.055
76. Mangan SD, Koçlar Oral G (2016) Life cycle assessment of energy retrofit strategies for an existing residential building in Turkey. A/Z ITU J Fac Archit 13(2):143–156. https://doi.org/10.5505/itujfa.2016.26928
77. Valančius K, Vilutienė T, Rogoža A (2018) Analysis of the payback of primary energy and CO_2 emissions in relation to the increase of thermal resistance of a building. Energy Build. 179:39–48. https://doi.org/10.1016/J.ENBUILD.2018.08.037
78. Mangan SD, Oral GK (2015) A study on life cycle assessment of energy retrofit strategies for residential buildings in Turkey. Energy Procedia 78:842–847. https://doi.org/10.1016/J.EGYPRO.2015.11.005
79. Marique A-F, Rossi B (2018) Cradle-to-grave life-cycle assessment within the built environment: comparison between the refurbishment and the complete reconstruction of an office building in Belgium. J Environ Manage 224:396–405. https://doi.org/10.1016/J.JENVMAN.2018.02.055
80. Oregi X, Hernandez P, Hernandez R (2017) Analysis of life-cycle boundaries for environmental and economic assessment of building energy refurbishment projects. Energy Build. 136:12–25. https://doi.org/10.1016/j.enbuild.2016.11.057

81. Nydahl H, Andersson S, Åstrand A, Olofsson T (2019) Environmental performance measures to assess building refurbishment from a life cycle perspective. Energies 12(2):299. https://doi.org/10.3390/en12020299
82. Oregi X, Hernandez P, Gazulla C, Isasa M (2015) Integrating simplified and full life cycle approaches in decision making for building energy refurbishment: benefits and barriers. Buildings 5(2):354–380. https://doi.org/10.3390/buildings5020354
83. Tadeu S, Rodrigues C, Tadeu A, Freire F, Simões N (2015) Energy retrofit of historic buildings: Environmental assessment of cost-optimal solutions. J Build Eng 4:167–176. https://doi.org/10.1016/J.JOBE.2015.09.009
84. Ouyang J, Lu M, Li B, Wang C, Hokao K (2011) Economic analysis of upgrading aging residential buildings in China based on dynamic energy consumption and energy price in a market economy. Energy Policy 39(9):4902–4910. https://doi.org/10.1016/J.ENPOL.2011.06.025
85. Toosi HA, Balador Z, Gjerde M, Vakili-Ardebili A (2018) A life cycle cost analysis and environmental assessment on the photovoltaic system in buildings: two case studies in Iran. J Clean Energy Technol 6(2):134–138. https://doi.org/10.18178/jocet.2018.6.2.448
86. Jafari A, Valentin V, Russell M (2016) Sensitivity analysis of factors affecting decision-making for a housing energy retrofit: a case study, no. May 2016, pp 1254–1263. https://doi.org/10.1061/9780784479827.126
87. Copiello S, Gabrielli L, Bonifaci P (2017) Evaluation of energy retrofit in buildings under conditions of uncertainty: the prominence of the discount rate. Energy 137:104–117. https://doi.org/10.1016/j.energy.2017.06.159
88. Lucchi E, Tabak M, Troi A (2017) The 'cost optimality' approach for the internal insulation of historic buildings. Energy Procedia 133:412–423. https://doi.org/10.1016/J.EGYPRO.2017.09.372
89. Yuan J, Nian V, Su B (2019) Evaluation of cost-effective building retrofit strategies through soft-linking a metamodel-based Bayesian method and a life cycle cost assessment method. Appl Energy 253: 113573, 2019. https://doi.org/10.1016/j.apenergy.2019.113573
90. Liu L, Rohdin P, Moshfegh B (2016) LCC assessments and environmental impacts on the energy renovation of a multi-family building from the 1890s. Energy Build. 133:823–833. https://doi.org/10.1016/J.ENBUILD.2016.10.040
91. Copiello S, Gabrielli L, Bonifaci P (2017) Evaluation of energy retrofit in buildings under conditions of uncertainty: the prominence of the discount rate. Energy 137:104–117, Oct. 2017. https://doi.org/10.1016/J.ENERGY.2017.06.159
92. Bleyl JW et al (2019) Office building deep energy retrofit: life cycle cost benefit analyses using cash flow analysis and multiple benefits on project level. Energy Effic 12(1):261–279. https://doi.org/10.1007/s12053-018-9707-8
93. Van De Moortel E et al (2019) Energy Renovation of Social Housing : Finding a Balance Between Increasing Insulation and Improving Heating System Efficiency Energy Renovation of Social Housing : Finding a Balance Between Increasing Insulation and Improving Heating System Efficiency," IOP Conf. Ser. Earth Environ Sci Cent Eur Towar Sustain Build. doi: https://doi.org/10.1088/1755-1315/290/1/012137
94. Ruparathna R, Hewage K, Sadiq R (2017) Economic evaluation of building energy retrofits: a fuzzy based approach. Energy Build. 139:395–406. https://doi.org/10.1016/j.enbuild.2017.01.031
95. Zheng D, Yu L, Wang L, Tao J (2019) A screening methodology for building multiple energy retrofit measures package considering economic and risk aspects. J Clean Prod 208:1587–1602. https://doi.org/10.1016/J.JCLEPRO.2018.10.196
96. Lohse R, Staller H, Riel M (2016) The economic challenges of deep energy renovation—differences, similarities, and possible solutions in central Europe: Austria and Germany. ASHRAE Conf. 122:69–87
97. Fregonara E, Lo Verso VRM, Lisa M, Callegari G (2017) Retrofit scenarios and economic sustainability. A case-study in the italian context. Energy Procedia, vol. 111, pp. 245–255, Mar. 2017, doi: https://doi.org/10.1016/J.EGYPRO.2017.03.026.

98. D'Orazio M, Di Giuseppe E, Esposti R, Coderoni S, Baldoni E (2018) A probabilistic tool for evaluating the effectiveness of financial measures to support the energy improvements of existing buildings.In: IOP Conf. Ser. Mater. Sci. Eng., vol 415, p 012003, Nov. 2018https://doi.org/10.1088/1757-899X/415/1/012003
99. Seghezzi R-CE, Masera G (2019) Decision Support for existing buildings : an LCC-based proposal for facade retrofitting technological choices Decision Support for existing buildings : an LCC-based proposal for facade retrofitting technological choices. In: IOP Conf. Ser. Earth Environ Sci 296, 2019, doi: https://doi.org/10.1088/1755-1315/296/1/012032
100. Krarti M, Dubey K (2017) Energy productivity evaluation of large scale building energy efficiency programs for Oman. Sustain Cities Soc 29:12–22. https://doi.org/10.1016/J.SCS.2016.11.009
101. Brás A, Rocha A, Faustino P (2015) Integrated approach for school buildings rehabilitation in a Portuguese city and analysis of suitable third party financing solutions in EU. J. Build. Eng. 3:79–93. https://doi.org/10.1016/J.JOBE.2015.05.003
102. Gustafsson S-I, Karlsson BG (1989) Insulation and bivalent heating system optimization: Residential housing retrofits and time-of-use tariffs for electricity. Appl Energy 34(4):303–315. https://doi.org/10.1016/0306-2619(89)90035-4
103. S. Gustafsson, "Energy conservation and optimal retrofits in multifamily buildings," Energy Syst. Policy, ISSN 0090–8347, Vol. 14, p. 37–49, pp. 1–13, 1990.
104. Gustafsson S-I, Karlsson BG (1991) Window retrofits: Interaction and life-cycle costing. Appl Energy 39(1):21–29. https://doi.org/10.1016/0306-2619(91)90060-B
105. Gustafsson S-I, Andersson S, Karlsson BG (1994) Factorial design for energy-system models. Energy 19(8):905–910. https://doi.org/10.1016/0360-5442(94)90043-4
106. Gustafsson S-I (1998) Sensitivity analysis of building energy retrofits. Appl Energy 61(1):13–23. https://doi.org/10.1016/S0306-2619(98)00032-4
107. V. Mili, K. Ekelöw, and B. Moshfegh, "On the performance of LCC optimization software OPERA-MILP by comparison with building energy simulation software IDA ICE," vol. 128, no. July 2017, pp. 305–319, 2018, doi: https://doi.org/10.1016/j.buildenv.2017.11.012.
108. C. Nägeli, A. Farahani, M. Österbring, J. Dalenbäck, and H. Wallbaum, "A service-life cycle approach to maintenance and energy retrofit planning for building portfolios," Build. Environ., vol. 160, no. June, p. 106212, 2019, doi: https://doi.org/10.1016/j.buildenv.2019.106212.
109. La Fleur L, Rohdin P, Moshfegh B (2019) Energy Renovation versus Demolition and Construction of a New Building — A Comparative Analysis of a Swedish Multi-Family Building. Energies 12:12–15. https://doi.org/10.3390/en12112218
110. Koo C, Kim H, Hong T (2014) Framework for the analysis of the low-carbon scenario 2020 to achieve the national carbon Emissions reduction target: Focused on educational facilities. Energy Policy 73:356–367. https://doi.org/10.1016/j.enpol.2014.05.009
111. Koo C, Hong T, Kim J, Kim H (2015) An integrated multi-objective optimization model for establishing the low-carbon scenario 2020 to achieve the national carbon emissions reduction target for residential buildings. Renew Sustain Energy Rev 49:410–425. https://doi.org/10.1016/J.RSER.2015.04.120
112. S. D. Mangan and G. Koçlar Oral, "A study on determining the optimal energy retrofit strategies for an existing residential building in Turkey," A/Z ITU J. Fac. Archit., vol. 11, no. 2, pp. 307–333, 2014.
113. E. S. Umdu, "Methodological approach for performance assessment of historical buildings based on seismic , energy and cost performance : A Mediterranean," J. Build. Eng., vol. 31, no. March, 2020, doi: https://doi.org/10.1016/j.jobe.2020.101372.
114. Y. Yılmaz and G. Koçlar Oral, "An approach for an educational building stock energy retrofits through life-cycle cost optimization," Archit. Sci. Rev., vol. 61, no. 3, pp. 122–132, May 2018, doi: https://doi.org/10.1080/00038628.2018.1447438.
115. Jafari A, Valentin V (2015) Decision-making life-cycle cost analysis model for energy-efficient housing retrofits. Int J Sustain Build Technol Urban Dev 6(3):173–187. https://doi.org/10.1080/2093761X.2015.1074948

116. A. Jafari and V. Valentin, "An Investment Allocation Approach for Building Energy Retrofits," no. May 2016, pp. 1061–1070, 2016, doi: https://doi.org/10.1061/9780784479827.107.
117. Mostavi E, Asadi S, Boussaa D (2017) Development of a new methodology to optimize building life cycle cost, environmental impacts, and occupant satisfaction. Energy 121:606–615. https://doi.org/10.1016/J.ENERGY.2017.01.049
118. Kušar M, Šubic M, Šelih J (2013) Selection of Efficient Retrofit Scenarios for Public Buildings. Procedia Eng. 57:651–656. https://doi.org/10.1016/j.proeng.2013.04.082
119. Holopainen R, Milandru A, Ahvenniemi H, Häkkinen T (2016) Feasibility Studies of Energy Retrofits – Case Studies of Nearly Zero-energy Building Renovation. Energy Procedia 96:146–157. https://doi.org/10.1016/J.EGYPRO.2016.09.116
120. Assiego de Larriva R, Calleja Rodríguez G, Cejudo López JM, Raugei M, Fullana i Palmer P (2014) A decision-making LCA for energy refurbishment of buildings: conditions of comfort," Energy Build 70:333–342, Feb. 2014. https://doi.org/10.1016/J.ENBUILD.2013.11.049
121. Sala S, Farioli F, Zamagni A (2013) Progress in sustainability science : lessons learnt from current methodologies for sustainability assessment : Part 1," pp 1653–1672. https://doi.org/10.1007/s11367-012-0508-6.
122. C. Llatas, B. Soust-verdaguer, and A. Passer, "Implementing Life Cycle Sustainability Assessment during design stages in Building Information Modelling : From systematic literature review to a methodological approach," Build. Environ., vol. 182, no. July, p. 107164, 2020, doi: https://doi.org/10.1016/j.buildenv.2020.107164.
123. Messerli P et al (2019) Expansion of sustainability science needed for the SDGs. Nat. Sustain. 2(10):892–894. https://doi.org/10.1038/s41893-019-0394-z
124. Pauliuk S (2020) Making sustainability science a cumulative effort. Nat. Sustain. 3(1):2–4. https://doi.org/10.1038/s41893-019-0443-7
125. J. Fariña-tojo and J. Rajaniemi, Urban Ecology, Emerging Patterns and Social-Ecological Systems, (Chapter 19 - Challenges in assessing urban sustainability), no. July. Elsevier Inc., 2020.
126. Sala S, Ciuffo B, Nijkamp P (2015) A systemic framework for sustainability assessment. Ecol Econ 119:314–325. https://doi.org/10.1016/j.ecolecon.2015.09.015
127. A. Zamagni, H. Pesonen, and T. Swarr, "From LCA to Life Cycle Sustainability Assessment : concept , practice and future directions," pp. 1637–1641, 2013, doi: https://doi.org/10.1007/s11367-013-0648-3.
128. A. Lehmann, E. Zschieschang, and M. Traverso, "Social aspects for sustainability assessment of technologies — challenges for social life cycle assessment (SLCA)," pp. 1581–1592, 2013, doi: https://doi.org/10.1007/s11367-013-0594-0.
129. I. Huertas-valdivia, A. M. Ferrari, D. Settembre-blundo, and F. E. Garc, "Social Life-Cycle Assessment : A Review by Bibliometric Analysis," pp. 1–25, 2020.
130. Contestabile M (2020) Measuring for sustainability. Nat. Sustain. 3(8):576. https://doi.org/10.1038/s41893-020-0570-1
131. D. S. Zachary, "On the sustainability of an activity," Sci. Rep., vol. 4, 2014, doi: https://doi.org/10.1038/srep05215.
132. Moallemi EA et al (2020) Perspective Achieving the Sustainable Development Goals Requires Transdisciplinary Innovation at the Local Scale. One Earth 3(3):300–313. https://doi.org/10.1016/j.oneear.2020.08.006
133. O. Tokede and M. Traverso, "Implementing the guidelines for social life cycle assessment : past , present , and future," pp. 1910–1929, 2020.
134. R. J. Bonilla-alicea and K. Fu, "Systematic Map of the Social Impact Assessment Field," 2019.
135. Alexander SM et al (2020) Qualitative data sharing and synthesis for sustainability science. Nat. Sustain. 3(2):81–88. https://doi.org/10.1038/s41893-019-0434-8
136. A. Zamagni, "Life cycle sustainability assessment," pp. 373–376, 2012, doi: https://doi.org/10.1007/s11367-012-0389-8.
137. Matthews NE, Stamford L, Shapira P (2019) Aligning sustainability assessment with responsible research and innovation : Towards a framework for Constructive Sustainability Assessment. Sustain. Prod. Consum. 20:58–73. https://doi.org/10.1016/j.spc.2019.05.002

138. J. Pedro, C. Silva, and M. Duarte, "Scaling up LEED-ND sustainability assessment from the neighborhood towards the city scale with the support of GIS modeling : Lisbon case study," Sustain. Cities Soc., vol. 41, no. May 2017, pp. 929–939, 2018, doi: https://doi.org/10.1016/j.scs.2017.09.015.
139. H. H. Dang and U. Serajuddin, "Tracking the sustainable development goals: Emerging measurement challenges and further reflections," World Dev., vol. 127, p. 104570, 2020https://doi.org/10.1016/j.worlddev.2019.05.024
140. R. Phillips, L. Troup, D. Fannon, and M. J. Eckelman, "Triple bottom line sustainability assessment of window-to-wall ratio in US office buildings," Build. Environ., vol. 182, no. January, p. 107057, 2020, doi: https://doi.org/10.1016/j.buildenv.2020.107057.
141. C. Galvão and W. Viegas, "Inquiry in higher education for sustainable development : crossing disciplinary knowledge boundaries," vol. 2020, 2020, doi: https://doi.org/10.1108/IJSHE-02-2020-0068.
142. S. Huysveld, S. E. Taelman, S. Sfez, and J. Dewulf, "A framework for using the handprint concept in attributional life cycle (sustainability) assessment," vol. 265, pp. 1–9, 2020, doi: https://doi.org/10.1016/j.jclepro.2020.121743.
143. Partelow S (2016) Coevolving Ostrom ' s social – ecological systems (SES) framework and sustainability science : four key co-benefits. Sustain Sci 11(3):399–410. https://doi.org/10.1007/s11625-015-0351-3
144. L. Wan, E. Ng, L. Wan, and E. Ng, "Evaluation of the social dimension of sustainability in the built environment in poor rural areas of China Evaluation of the social dimension of sustainability in the built environment in poor rural areas of China," vol. 8628, 2018, doi: https://doi.org/10.1080/00038628.2018.1505595.
145. P. Verma and A. S. Raghubanshi, "Urban sustainability indicators : Challenges and opportunities," vol. 93, no. May, pp. 282–291, 2018, doi: https://doi.org/10.1016/j.ecolind.2018.05.007.
146. Andreas R, Serenella S, Jungbluth N (2020) Normalization and weighting: the open challenge in LCA. Int J Life Cycle Assess. https://doi.org/10.1007/s11367-020-01790-0
147. N. Pelletier, N. Bamber, and M. Brandão, "Interpreting life cycle assessment results for integrated sustainability decision support : can an ecological economic perspective help us to connect the dots ?," pp. 1580–1586, 2019.
148. Pizzol M, Laurent A, Sala S, Weidema B, Verones F, Koffler C (2017) Normalisation and weighting in life cycle assessment: quo vadis? Int J Life Cycle Assess 22(6):853–866. https://doi.org/10.1007/s11367-016-1199-1
149. M. Abubakr, A. T. Abbas, I. Tomaz, M. S. Soliman, M. Luqman, and H. Hegab, "Sustainable and Smart Manufacturing : An Integrated Approach," 2020.
150. P. Halla, V. Superti, A. Boesch, and C. R. Binder, "Indicators for urban sustainability : Key lessons from a systematic analysis of 67 measurement initiatives," Ecol. Indic., vol. 119, no. April, p. 106879, 2020, doi: https://doi.org/10.1016/j.ecolind.2020.106879.
151. Krarti M, Dubey K (2018) Review analysis of economic and environmental benefits of improving energy efficiency for UAE building stock. Renew Sustain Energy Rev 82:14–24. https://doi.org/10.1016/J.RSER.2017.09.013
152. Ruparathna R, Hewage K, Sadiq R (2017) Rethinking investment planning and optimizing net zero emission buildings. Clean Technol Environ Policy 19(6):1711–1724. https://doi.org/10.1007/s10098-017-1359-4
153. H. Amini Toosi and M. Lavagna, "Life Cycle Sustainability Assessment (LCSA) and Optimization Techniques. A conceptual framework for integrating LCSA into designing energy retrofit scenarios of existing buildings," in The 12th Italian LCA network conference. Life cycle thinking in decision making for sustainability: from public policies to private business. University of Messina, Italy 11–12 JUNE 2018, 2018, pp. 276–284, doi: ISBN: 978–88–8286–372–2.
154. H. Amini Toosi and M. Lavagna, "Optimization and LCSA-based design method for energy retrofitting of existing buildings," in Designing Sustainability for All, Proceedings of the 3rd LeNS World Distributed Conference, Milano, Mexico City, Beijing, Bangalore, Curitiba, Cape Town, 3–5 April 2019, VOL.1, 2019, pp. 1107–1111, doi: ISBN: 978–88–95651–26–2.

155. V. Machairas, A. Tsangrassoulis, and K. Axarli, "Algorithms for optimization of building design : A review," vol. 31, no. 1364, pp. 101–112, 2014, doi: https://doi.org/10.1016/j.rser.2013.11.036.
156. Chantrelle FP, Lahmidi H, Keilholz W, El Mankibi M, Michel P (2011) Development of a multicriteria tool for optimizing the renovation of buildings. Appl Energy 88(4):1386–1394. https://doi.org/10.1016/J.APENERGY.2010.10.002
157. Risholt B, Time B, Grete A (2013) Sustainability assessment of nearly zero energy renovation of dwellings based on energy, economy and home quality indicators. Energy Build. 60:217–224. https://doi.org/10.1016/j.enbuild.2012.12.017
158. Gustafsson M, Swing M, Are J, Bales C, Holmberg S (2016) Techno-economic analysis of energy renovation measures for a district heated multi-family house. Appl Energy 177:108–116. https://doi.org/10.1016/j.apenergy.2016.05.104
159. Pal SK, Takano A, Alanne K, Siren K (2017) A life cycle approach to optimizing carbon footprint and costs of a residential building. Build Environ 123:146–162. https://doi.org/10.1016/j.buildenv.2017.06.051
160. Ramin H, Hanafizadeh P, Ehterami T, AkhavanBehabadi MA (2017) Life cycle-based multi-objective optimization of wall structures in climate of Tehran. Adv. Build. Energy Res. 2549:1–14. https://doi.org/10.1080/17512549.2017.1344137
161. Moschetti R, Brattebø H (2017) Combining Life Cycle Environmental and Economic Assessments in Building Energy Renovation Projects. Energies. https://doi.org/10.3390/en10111851
162. Ylmén P, Mjörnell K, Berlin J, Arfvidsson J (2017) The influence of secondary effects on global warming and cost optimization of insulation in the building envelope. Build Environ 118:174–183. https://doi.org/10.1016/J.BUILDENV.2017.03.019
163. Gustafsson M et al (2017) Economic and environmental analysis of energy renovation packages for European office buildings. Energy Build. 148:155–165. https://doi.org/10.1016/j.enbuild.2017.04.079
164. Mauro G et al (2017) A Multi-Step Approach to Assess the Lifecycle Economic Impact of Seismic Risk on Optimal Energy Retrofit. Sustainability 9(6):989. https://doi.org/10.3390/su9060989
165. Almeida M, Ferreira M (2017) Cost effective energy and carbon emissions optimization in building renovation (Annex 56). Energy Build. 152:718–738. https://doi.org/10.1016/j.enbuild.2017.07.050
166. M. Almeida, "Relevance of Embodied Energy and Carbon Emissions on Assessing Cost Effectiveness in Building Renovation — Contribution from the Analysis of Case Studies in Six European Countries," pp. 1–18, 2018, doi: https://doi.org/10.3390/buildings8080103.
167. J. Jokisalo, P. Sankelo, J. Vinha, K. Sirén, and R. Kosonen, "Cost optimal energy performance renovation measures in a municipal service building in a cold climate," in E3S Web of Conferences 111, CLIMA 2019, 2019, vol. 3022, no. 201 9, doi: https://doi.org/10.1051/e3sconf/201911103022.
168. S. Amirhosain and A. Hammad, "Developing surrogate ANN for selecting near-optimal building energy renovation methods considering energy consumption , LCC and LCA," J. Build. Eng., vol. 25, no. April, p. 100790, 2019, doi: https://doi.org/10.1016/j.jobe.2019.100790.
169. J. Hirvonen, J. Jokisalo, J. Heljo, and R. Kosonen, "Optimization of emission reducing energy retrofits in Finnish apartment buildings," E3S Web Conf. 111, CLIMA 2019, vol. 2, no. 2019, 2019, doi: https://doi.org/10.1051/e3sconf/2019111030.
170. Conci M, Konstantinou T, Van Den Dobbelsteen A, Schneider J (2019) Trade-off between the economic and environmental impact of different decarbonisation strategies for residential buildings. Build Environ 155(January):137–144. https://doi.org/10.1016/j.buildenv.2019.03.051
171. R. Mateus, S. M. Silva, and M. G. De Almeida, "Environmental and cost life cycle analysis of the impact of using solar systems in energy renovation of Southern European single-family buildings," Renew. Energy, vol. 137, pp. 82–92, 2019, doi: https://doi.org/10.1016/j.renene.2018.04.036.

172. A. Dalla Valle, A. Campioli, and M. Lavagna, "Life cycle BIM-oriented data collection: A framework for supporting practitioners," in Research for Development, Springer, 2020, pp. 49–59.
173. Lu Y, Wu Z, Chang R, Li Y (2017) Building Information Modeling (BIM) for green buildings: A critical review and future directions. Autom Constr 83:134–148. https://doi.org/10.1016/J.AUTCON.2017.08.024
174. Najjar M, Figueiredo K, Palumbo M, Haddad A (2017) Integration of BIM and LCA: Evaluating the environmental impacts of building materials at an early stage of designing a typical office building. J. Build. Eng. 14:115–126. https://doi.org/10.1016/J.JOBE.2017.10.005
175. Shadram F, Johansson TD, Lu W, Schade J, Olofsson T (2016) An integrated BIM-based framework for minimizing embodied energy during building design. Energy Build. 128:592–604. https://doi.org/10.1016/j.enbuild.2016.07.007
176. Hollberg A, Ruth J (2016) LCA in architectural design—a parametric approach. Int J Life Cycle Assess 21(7):943–960. https://doi.org/10.1007/s11367-016-1065-1
177. Antón LÁ, Díaz J (2014) Integration of life cycle assessment in a BIM environment. Procedia Eng. 85:26–32. https://doi.org/10.1016/j.proeng.2014.10.525
178. Eleftheriadis S, Mumovic D, Greening P (2017) Life cycle energy efficiency in building structures: A review of current developments and future outlooks based on BIM capabilities. Renew Sustain Energy Rev 67:811–825. https://doi.org/10.1016/J.RSER.2016.09.028
179. Barlish K, Sullivan K (2012) How to measure the benefits of BIM — A case study approach. Autom Constr 24:149–159. https://doi.org/10.1016/J.AUTCON.2012.02.008
180. Bryde D, Broquetas M, Volm JM (2013) The project benefits of Building Information Modelling (BIM). Int J Proj Manag 31(7):971–980. https://doi.org/10.1016/J.IJPROMAN.2012.12.001
181. Kota S, Haberl JS, Clayton MJ, Yan W (2014) Building Information Modeling (BIM)-based daylighting simulation and analysis. Energy Build. 81:391–403. https://doi.org/10.1016/J.ENBUILD.2014.06.043
182. Soust-Verdaguer B, Llatas C, García-Martínez A (2017) Critical review of bim-based LCA method to buildings. Energy Build. 136:110–120. https://doi.org/10.1016/J.ENBUILD.2016.12.009
183. Malmqvist T et al (2011) Life cycle assessment in buildings: The ENSLIC simplified method and guidelines. Energy 36(4):1900–1907. https://doi.org/10.1016/J.ENERGY.2010.03.026
184. Peng C (2016) Calculation of a building's life cycle carbon emissions based on Ecotect and building information modeling. J Clean Prod 112:453–465. https://doi.org/10.1016/J.JCLEPRO.2015.08.078
185. N. Shafiq, M. F. Nurrudin, S. S. S. Gardezi, and A. Bin Kamaruzzaman, "Carbon footprint assessment of a typical low rise office building in Malaysia using building information modelling (BIM)," Int. J. Sustain. Build. Technol. Urban Dev., vol. 6, no. 3, pp. 157–172, Jul. 2015, doi: https://doi.org/10.1080/2093761X.2015.1057876.
186. Basbagill J, Flager F, Lepech M, Fischer M (2013) Application of life-cycle assessment to early stage building design for reduced embodied environmental impacts. Build Environ 60:81–92. https://doi.org/10.1016/J.BUILDENV.2012.11.009
187. Ajayi SO, Oyedele LO, Ceranic B, Gallanagh M, Kadiri KO (2015) Life cycle environmental performance of material specification: a BIM-enhanced comparative assessment. Int J Sustain Build Technol Urban Dev 6(1):14–24. https://doi.org/10.1080/2093761X.2015.1006708
188. Jalaei F, Jrade A (2014) An Automated BIM Model to Conceptually Design, Analyze, Simulate, and Assess Sustainable Building Projects. J Constr Eng 2014:1–21. https://doi.org/10.1155/2014/672896
189. Lee S, Tae S, Roh S, Kim T (2015) Green Template for Life Cycle Assessment of Buildings Based on Building Information Modeling: Focus on Embodied Environmental Impact. Sustainability 7(12):16498–16512. https://doi.org/10.3390/su71215830
190. J. Basbagill, F. Flager, M. Lepech, and M. Fischer, "Application of life-cycle assessment to early stage building design for reduced embodied environmental impacts," Build. Environ., vol. 60, pp. 81–92, Feb. 2013, doi: https://doi.org/10.1016/J.BUILDENV.2012.11.009.

191. A. Houlihan Wiberg et al., "A net zero emission concept analysis of a single-family house," Energy Build., vol. 74, pp. 101–110, May 2014, doi: https://doi.org/10.1016/J.ENBUILD.2014.01.037.
192. B. Berg, "Using Bim To Calculate Accurate Building Material Quantities for Early Design Phase Life Cycle Assessment," 2014.
193. Dupuis M, April A, Lesage P, Forgues D (2017) Method to enable LCA analysis through each level of development of a BIM model. Procedia Eng. 196:857–863. https://doi.org/10.1016/J.PROENG.2017.08.017

Evaluating the Sustainability of Feedlot Production in Australia Using a Life Cycle Sustainability Assessment Framework

Murilo Pagotto, Anthony Halog, Diogo Fleury Azevedo Costa, and Tianchu Lu

Abstract This chapter presents and discusses the results of a case study completed to evaluate and develop strategies to improve the sustainability of beef production using the feedlot system in Australia. The study was developed to test a proposed sustainability assessment framework that uses Life Cycle Sustainability Assessment (LCSA) methodologies to assess the sustainability of beef production using a feedlot system in the central region of the state of Queensland, Australia. Beef production was selected because the sector is strongly linked to climate change and other environmental and socio-economic impacts worldwide. Despite being a sector considered environmentally unsustainable when the correct agronomic practices are not in place, it has fundamental socio-economic importance to the country, especially in remote rural communities. Thus, beyond this analysis, this study also reports the application of the proposed sustainability assessment methodology to model and analyse different scenarios potentially created by the implementation of sustainable technologies and circular economy principles in cattle feedlot production systems in Australia. Following these ideologies, the study presented in this chapter assessed the sustainability of a feedlot beef production linear model and considered how the implementation of sustainable approaches that use the principles of circularity and sustainable development would affect the overall sustainability of this complex system. The assessment was performed using a proposed sustainability assessment framework designed to evaluate the sustainability of food systems and verify the effects of the implementation of sustainable production processes in the system. The framework uses LCSA techniques and modelling to evaluate how resources are extracted, processed, consumed and disposed in the natural environment as well as holistic measures of the sustainability and efficiency of complex systems. Additionally, the proposed framework could be used to appraise the consumption and

---

M. Pagotto (✉) · A. Halog · T. Lu
School of Earth and Environmental Science, University of Queensland, Brisbane St Lucia, QLD 4072, Australia
e-mail: murilopagotto@live.com

D. F. A. Costa
Institute for Future Farming Systems, Central Queensland University, Rockhampton, QLD 4701, Australia

S. S. Muthu (ed.), *Life Cycle Sustainability Assessment (LCSA)*,
Environmental Footprints and Eco-design of Products and Processes,
https://doi.org/10.1007/978-981-16-4562-4_7

production patterns of a particular economy or region to demonstrate how that society utilises the available resources to satisfy its needs in a sustainable or unsustainable manner.

Keywords Life cycle sustainability assessment · Sustainability · Resource efficiency · Sustainable production and consumption · Material circularity · Feedlot production

1 Introduction

The demand for protein sources, such as beef, has exponentially increased in the last decade, especially in developing countries. To supply this growing demand, the number of beef producers adopting feedlot systems has also increased [41, 51]. Last year alone, the Australian feedlot industry reached record numbers of approximately 3.4 million head-on confined systems, with slaughters reaching 38% as 'grain-fed' cattle (MLA 2020). Feedlot systems are highly mechanised and integrated, they are more efficient and the resulting body weight gain (BWG) is significantly higher than those of grazing and crop and livestock integrated systems [41]. Although feedlot systems have production efficiency advantages, the risk of increased pollution created by their intensive production model is greater if a system is poorly managed [41]. Generally, air pollution, the risk of water contamination and high natural resource use are the major deleterious effects of the implementation of feedlot systems [41].

The Australian beef industry is the largest agricultural sector in the country and generates 3% of the global beef production [66]. The Australian cattle herd reached a staggering number of 26.4 million heads in 2019 (MLA 2020). Thus, Australia is one of the largest beef producers and live cattle exporters in the world [87]. Australian beef is exported to over 100 different international markets and benefits from the Australian reputation as a producer of high-quality food, particularly in Asian markets [87]. The state of Queensland is the largest beef production region in Australia and accounts for almost half of the country's national herd. The beef production system in Australia is highly diversified and employs different cattle production systems, including the popular cattle feedlot system. In Australia, there are more than 450 accredited beef feedlot enterprises and the cattle feedlot industry produces around 80% of beef sold in the domestic market as well as exports high-quality beef to international markets [87].

The Australian beef production sector is both economically and socially important (Table 1). Beef production is part of the Australian livestock industry, which directly and indirectly employed more than 430,000 Australian workers and generated AUD$62 billion during 2015–16 [66]. The industry has more than 43,000 registered commercial and family-operated businesses, which directly employ more than 77,000 Australian workers and provide income for thousands of farmers in regional areas [87].

Table 1 Key statistics of the Australian beef cattle farming industry

Period	Revenue (AUD$m)	Enterprises (units)	Employment (units)	Exports ($m)	Wages ($m)	Domestic demand ($m)
2009–10	14,286.3	47,110	81,331	809.0	1,142.2	13,477.4
2010–11	15,219.5	46,710	78,843	750.4	1,048.6	14,469.2
2011–12	13,319.5	46,317	73,731	693.3	869.0	12,626.3
2012–13	11,838.0	45,734	72,807	627.9	807.6	11,210.2
2013–14	12,785.5	45,940	76,644	1,101.7	889.3	11,683.9
2014–15	14,900.2	44,780	76,412	1,434.6	909.4	13,465.7
2015–16	18,617.4	44,632	77,870	1,648.6	1,121.7	16,968.9
2016–17	19,096.2	44,419	77,719	1,229.0	1,100.0	17,867.3
2017–18	17,259.3	44,107	76,754	1,282.6	1,095.8	15,976.8
2018–19	17,007.9	43,308	77,335	1,290.6	1,065.2	15,717.4
2019–20	17,194.3	43,390	76,902	1,330.8	1,077.6	15,863.6
2020–21	17,133.2	43,330	76,979	1,391.7	1,103.2	15,741.6
2021–22	17,904.2	43,144	76,176	1,366.2	1,113.5	16,538.1
2022–23	18,459.2	43,004	75,957	1,405.3	1,132.7	17,054.0
2023–24	18,883.8	42,954	75,964	1,424.6	1,140.7	17,459.3

Source [87]

While the beef industry is important to the Australian economy and society, it causes environmental degradation [3]. Its supply chain creates environmental impacts throughout the production life cycle, and GHG emissions and pollutants that affect aquatic and terrestrial ecosystems are generated during the production and processing of meat products [3]. Additionally, beef production requires large amounts of natural resources, particularly primary energy and water [75, 103]. The sector is one of the largest consumers of water within the Australian food system and requires considerable amounts of primary energy during its operation [75].

Understanding the economic, social, cultural and environmental factors involved in livestock production systems and how they interact is fundamental. In addition, to increase the sustainability of conventional livestock production systems, the issues associated with these factors must be correctly evaluated and addressed [35, 111]. The utilisation of sustainability assessment methodologies appears to be an efficient approach to measure the trade-off between socio-economic importance and environmental matters [75].

Based on these evidences, it is important to evaluate the sustainability of beef production in feedlot systems and implement sustainable technologies to reduce the detrimental environmental impacts and increase the systems' overall sustainability. According to [42], approaches to increase the sustainability of beef production already exist, however, they are not widely used. The authors argued that global action involving all stakeholders in livestock production, including the general community and decision-makers, is required to implement cost-effective strategies to increase the sustainability of feedlot systems.

The key factors in assessing the sustainability of a beef production enterprise are natural resource stewardship, animal health and welfare, profitability and social responsibility [12, 47]. The Global Roundtable for Sustainable Beef (GRSB) defined sustainable beef as

> socially responsible, environmentally sound and economically viable product that prioritizes **Planet** (relevant principles: *Natural Resources, Efficiency & Innovation, People and the Community*); **People** (relevant principles: *People and the Community and Food*); **Animals** (relevant principle: *Animal health and welfare*); and **Progress** (relevant principles: *Natural Resources, People and the Community, Animal health and welfare, Food, Efficiency and Innovation*) [51],p. 5, 'emphasis added').

As mentioned previously in this chapter, the sustainability of beef production using a feedlot system in Australia was selected for evaluation using a proposed sustainability assessment framework. There were several reasons for the selection of this particular food production sector including the facts presented in the previous statements. Another reason for the selection of beef production in this case study was based on the results of a study completed by [75]. The authors evaluated the eco-efficiency performance of various sub-sectors in the Australian agri-food systems through the use of input–output-oriented approaches of Data Envelopment Analysis (DEA) and Material Flow Analysis (MFA). One of the main objectives of this study was to establish environmental and economic indicators for the industry and to verify inefficiencies during the life cycle of food production in Australia. The study revealed the potential of eco-efficiency performance measures in Australian agri-food industry sub-sectors. The MFA results demonstrate that beef cattle farming accounted for more than 24% of the total CO2eq emissions generated by the Australian food supply chain [75].

Table 2 shows the results of the DEA performed by [75] to identify inefficient sub-sectors of the Australian food system. Moreover, these results revealed important facts related to food production in Australia. They show the resource efficiency of the Australian food industry sub-sectors and the GHG emitted by them as well as they identified the connections between economic benefits and environmental efficiency of these particular sub-sectors [75]. The value of 1 (100%) represents the efficiency frontier based on the DEA calculations. Therefore, the sub-sectors that received the value of 1 (100%) are considered efficient in the DEA assessment performed in this study. Though it is critical to emphasise that the primary objective of DEA is to perform an eco-efficiency performance evaluation of the sub-sectors analysed in this assessment benchmarking them against each other [75]. Consequently, several sub-sectors were considered efficient in this DEA, but this does not mean that these sub-sectors do not create environmental issues and require sizeable quantities of natural resources during their production cycle. Rather, the DEA results are relative: they demonstrate which sub-sectors are more efficient than in managing their inputs and outputs increasing their resource efficiency and reducing waste and air contaminants emissions [75].

Table 2 Results of DEA of the Australian food supply chain

DMU no.	DMU name	Input-oriented CRS efficiency
Processed food industry sub-sectors		
1	Dairy products	0.44 (44%)
2	Sugar and confectionary	1.0 (100%)
3	Meat and meat products	1.0 (100%)
4	Other food products	0.44 (44%)
5	Grain mill and cereal products	0.44 (44%)
6	Beer	1.0 (100%)
7	Bakery products	1.0 (100%)
8	Wine, spirits and tobacco	1.0 (100%)
9	Fruit and vegetable products	0.42 (42%)
10	Soft drinks, cordials and syrup	1.0 (100%)
11	Oils and fats	0.16 (16%)
Farming sub-sectors		
1	Horticulture	1.0 (100%)
2	Grain growing	1.0 (100%)
3	Beef cattle farming	0.37 (37%)
4	Poultry meat and eggs farming	1.0 (100%)
5	Dairy cattle farming	0.74 (74%)
6	Sugarcane growing	1.0 (100%)
7	Rice growing	0.91 (91%)
8	Pig farming	1.0 (100%)
9	Sheep farming	0.62 (62%)

Source [75]. DMU = decision making unit; CRS = constant return to scale

According to the DEA results, beef cattle farming was considered the least efficient sub-sector of the Australian food industry in regards to its eco-efficiency performance. However, it is well known that the Australian livestock sector creates complex environmental issues, although it produces considerable incomes adding value to the Australian economy. To more accurately verify these evidences, the case study presented in this chapter complemented the research of [75] to holistically assess the sustainability of the beef sector in Australia and attempted to propose solutions to solve the inefficiencies verified by these complementary studies.

The last motive for the selection of beef production for sustainability analysis is that there are no studies that holistically assessed the sustainability of this important sector of the Australian food system. Previous studies only assessed one of the three pillars of sustainability using a life cycle-based approach or other methodologies. Based on these evidences, this study not only attempted to inclusively assess the sustainability of beef production in Australia but also used LCSA methods and CE principles to resolve sustainability issues verified during the sustainability analysis.

One of the main objectives of this chapter was to develop a case study to test a sustainability assessment framework proposed and presented in the chapter 2 of this book by Pagotto et al. [74] called 'Food Systems Sustainability Assessment Framework' (FSSAF) to holistically analyse the sustainability of feedlot production in Australia. Additionally, the case study attempts to identify the inefficient processes that are currently used during feedlot production in Australia and propose sustainable and circular methods and technologies to improve the industry's sustainability. Lastly, the FSSAF uses LCSA modelling approaches to uncover and evaluate the accompanying benefits and impacts of the proposed changes for the feedlot industry.

2 Case Study

2.1 Goal and Scope

To test the proposed sustainability assessment framework, a case study was developed to assess the sustainability of a cattle feedlot system model in Australia. The model was created using the parameters of a large beef cattle feedlot enterprise and simulated a beef producer in the Central Queensland (CQ) region that uses a cattle feedlot system with the capacity to finish 50,000 head of Black Angus per cycle (150 days) for the Japanese market. The data to construct the model were collected from peer-reviewed private and government reports, scientific articles and databases.

The proposed FSSAF was applied to holistically evaluate the sustainability of cattle beef feedlot production in Australia. The inventory data collection and analysis were conducted following the LCSA guidelines published by UNEP, which cover the three life cycle analysis techniques included in this type of assessment [92]. The estimation and assessment of the environmental and socio-economic impacts as well as the possible positive and negative impacts created by the implementation of sustainable technologies into the feedlot system production model were calculated using the modelling tool OpenLCA 1.7.0. beta [73].

2.2 System Boundaries and Functional Unit

Each life cycle-based approach (LCA, LCC and SLCA) requires different system boundaries, particularly when applied to assess the sustainability of products and production systems [92, 110]. When performing LCSA, it is suggested that the system boundaries include all production processes of at least one of the life cycle-based approaches [92]. Following this principle, the LCSA conducted in this case study analysed the main gate-to-gate processes relevant to environmental life cycle analysis in a common beef cattle feedlot production system. The contribution of cattle breeding and backgrounding upstream processes was intentionally excluded from the assessment, since the study intended to conduct a gate-to-gate LCSA of cattle finishing using a feedlot system. However, the LCSA conducted a pre-farm analysis of the impacts generated by the upstream processes involved in the production of inputs used during the operation of a cattle feedlot production system. The environmental and socio-economic impacts of these processes were included in the LCSA assessment performed in the case study.

The main characteristics of the beef cattle feedlot model assessed by the FSSAF are the following: animals (Black Angus Breed); water and energy use; feed inputs; veterinary products; animal and input transportation; manure and waste management and mineral supplements. Further explanations of the characteristics of the developed model are presented in the following sections. The functional unit of the LCSA was one kilogram of beef BWG at the feedlot gate.

2.3 *Model Description*

The state of Queensland is the largest cattle producer in Australia, accounting for more than 41% of the total cattle production in the country [66]. According to [66], during the 2015–16 financial years, the Queensland cattle herd reached approximately 11.3 million head. Additionally, during this period, the state had the largest feedlot capacity in the country and it is estimated that more than 650,000 heads were finished in feedlot operations [66].

The Central Queensland (CQ) region was selected to simulate a representative feedlot operation to be analysed by the case study due to its location, climatic conditions and ability to accommodate a large number of large feedlot operators. Central Queensland covers more than 11.7 million hectares and has a sub-tropical climate with well-defined seasons and annual precipitation ranging from 600 to 900 mm [17]. Agriculture is the main economic activity in CQ and beef production uses 80% of the region's available arable land [17].

The proposed model designed in this case study was named Australian Cattle Feedlot (ACF) and represents a large cattle feedlot in CQ and its production processes (Fig. 1). The system was built using Australian beef feedlot production data from government research institutes, peer-reviewed studies and institutions involved in beef production research that support local beef producers. The case study simulated the representative enterprise as located near Rockhampton in CQ. The following sections provide more detailed information about the ACF model.

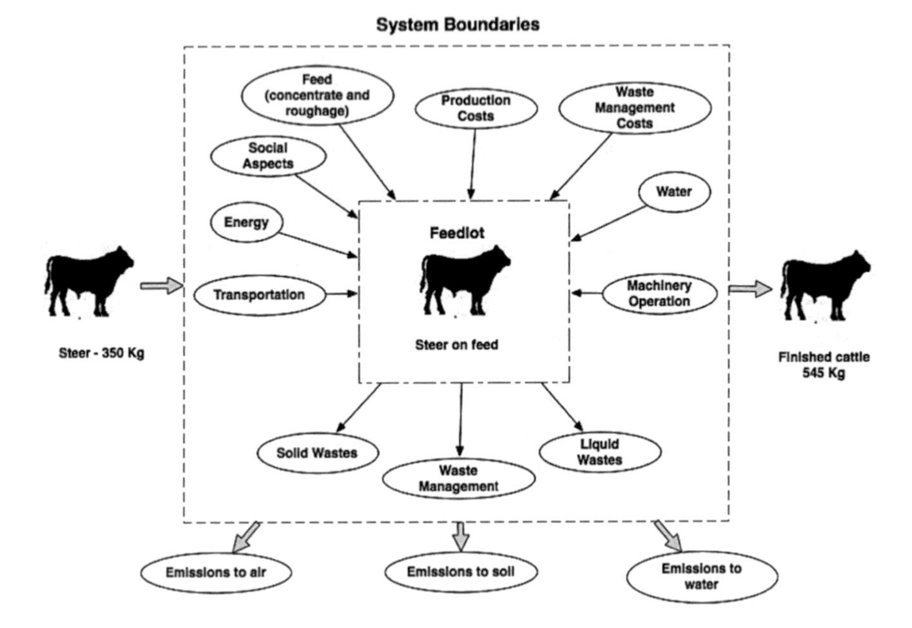

Fig. 1 Description of the Australian Cattle Feedlot (ACF) model

2.4 *Life Cycle Sustainability Assessment Inventory*

The case study used the principles and guidelines included in the FSSAF. The framework uses LCSA methodologies proposed by [88] and [91–93] to perform LCSA inventory analysis. To produce the environmental, economic and social inventories, the case study collected environmental and socio-economic data from peer-reviewed studies and reports (published by government agencies and private institutions). Following the methodologies of [88] and [91–93], the collected data were used to estimate the resource inputs, production costs, and social benefits and implications of the ACF model (Fig. 2).

2.4.1 Environmental Inventory

The environmental inventory of the ACF model was produced based on the resource input requirements of a large beef cattle feedlot operation in Australia. The main operational and production parameters of the simulated model are presented in Table 3.

The environmental inputs and outputs associated with the production processes of the ACF model were based on several studies and databases. Energy use (including fossil fuel) and electricity demand were estimated using data available from AusLCI, Ecoinvent 3.3 and peer-reviewed studies and reports (scientific studies) that evaluated

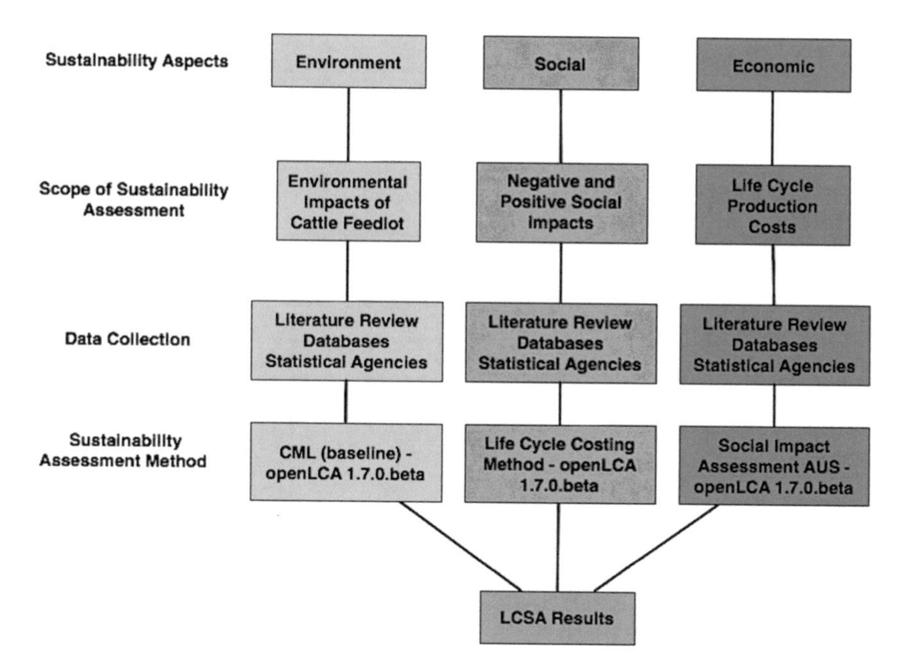

Fig. 2 LCSA method applied in this study. Data Source: [73]

Table 3 Characteristics of the ACF model

Parameters	Quantity	References
Capacity (head)	50,000	–
Hand feed-feedlot (days)	150	[64, 98, 105]
Feedlot total area (ha)	75	[64, 98, 105]
Average live weight entering the feedlot (kg)	350	[64, 98, 105]
Average live weight exiting the feedlot (kg)	625	[64, 98, 105]
Dressing weight (%)	56	[63, 66]
Dry matter feed intake (kg/head/day)	14.7	[40, 65, 105]
Average daily gain (kg/head/day)	1.4	[40, 65, 106]
Dry matter digestibility (%)	80	[40]
Carcass weight (%)	56	[66]
Dry matter content in manure (%)	66	[28]
Stock density (m^2)	9	[98]

beef feedlot production in Queensland and Australia. Large volumes of freshwater are required during feedlot operations. The direct ACF water consumption was based on data obtained from various reports and scientific studies (Table 4). Freshwater is mainly used in Australian feedlot operations for cattle drinking, cleaning activities

Table 4 Environmental inventory of the ACF model

Input flow	Unit	Amount/Head finished	References
Barley	kg	1420	Davis and Watts [29], [40], Wiedmann [105]
Diesel burned in electric generators	MJ	7.8	[38], Wiedmann et al. [105]
Dipping cattle (plunge dip)	item(s)	1	[64]
Electricity	MJ	1072.5	[38], Wiedmann et al. [105]
Mineral supplements	kg	136.5	[40, 65]
Natural gas QLD	MJ	204.7	[38], Wiedmann et al. [105]
Occupation, agriculture	m^2*a	11	[64, 65, 98]
Solid manure loading and spreading	kg	1189.5	[90]
Sorghum silage	kg	996	[40], Wiedmann et al. [105]
Tractor engine operations in Australia	kg	1.85	[90]
Transportation	t*km	300	Davis and Watts [29], [98]
Vaccination (5-in-1 vaccine)	item(s)	1	[25, 80]
Water Use	kg	14,235	Davis and Watts [29], [33, 79], Wiedmann et al. [105]

and feed preparation. In most cases, Australian feedlot operators extract fresh water from different sources such as rivers, dams, channels and artisan bores [98, 105].

Transportation activities involved in the ACF production processes were estimated using the data from [98], Davis and Watts [29] and Wiedmann et al. [105]. These data were analysed and used to build the transportation requirements of the ACF model. It was assumed that the model uses trucks with a carrying capacity of 28 tons for cattle and commodities transportation from distribution areas to the feedlot [29]. All other transportation activities (e.g., feed mixer loading, feed delivery to feedlot pens and manure management) were performed using tractors and tractor-drawn vertical mixers [98, 105].

Cattle nutrition and feed requirement parameters were estimated using the information and data from government organisations and scientific studies as referenced in Table 5. Veterinary products and mineral supplements parameters were estimated following the same procedure.

2.4.2 Manure Production

Manure is one of the main outputs of beef production using a feedlot system and the main product handled during waste management. Manure is normally considered a co-product of the beef system; it has commercial value and can be sold as fertiliser for other agricultural activities [105]. In this case study, the manure emissions of the ACF

Table 5 Production costs of the ACF model

Costs	AUD$/head	A$/kg	References
Purchase of store beast (310 kg at $1.40)	434	1.4	[7, 59]
Feed cost (1857 kg at $200 per ton)	360	1.84	[7, 49, 59]
Running costs	5	0.0025	[36, 59]
Cartage: saleyards to feedlot	6.5	0.033	[7]
Health cost: vaccines, drenches, etc	4.34	0.022	[59]
Losses: 1.0% of $434	3.73		[5, 59]
Fuel, repairs, etc	12	0.061	[59]
Labour	10	0.051	[5, 59]
Cartage: feedlot to abattoir	0		[7]
Selling costs (commission, etc.)	3.5	0.019	[49]
Transaction levy	7.83	0.04	[49]
Interest (feed + stock) 7% over 70 days	112.5	0.57	[8, 49]
Overhead costs (e.g., depreciation)	3.5	0.019	[8, 49]
Total	962.9	4.0575	[8, 49]

model were handled separately and calculated using the FarmGAS Calculator ST tool [7]. This is a decision-making tool, developed by the Australian Farm Institute that was designed to examine how different management systems and production practices might modify the GHG emission profile of a farm business or enterprise activity [8].

2.4.3 Production Costs Inventory

Life Cycle Costing assesses the total costs of a system or product throughout its life cycle using economic analysis [52]. The main objective of LCC analysis is to deliver an economic assessment of the processes and technologies used in the product system. This type of assessment could be useful to stakeholders involved in beef feedlot production in Australia to detect hidden costs and compare the trade-offs and benefits of the implementation of new and cleaner technologies for beef production [39, 52]. According to [52], LCC is a cost management method to estimate the costs associated with a product within the sustainability framework. It is an efficient tool to compare cost management alternatives for production systems.

The LCC inventory of the ACF model includes all the costs borne in the product's life cycle within the delineated system boundaries. It followed the LCC method proposed by [52]. Production costs data were collected from catalogues and data cubes published by statistical and economic government agencies in Australia as well as recently published peer-reviewed reports and scientific studies. The collected data were then allocated and normalised to express the cost of producing the functional unit in Australian dollars (Table 5).

2.4.4 Social Inventory

Social Life Cycle Assessment is a life cycle-based technique developed to investigate the social impacts created by production processes and services [57, 61, 91]. To build the SLCA inventory, this case study collected and analysed statistical data from different sources, including Australian Bureau of Statistics (ABS), International Labour Organization (ILO), Safe Work Australia and Public Health Information Development Unit (PHIDU). Additional statistical and qualitative data were collected from peer-reviewed scientific studies and Australian laws and regulations. The data collection and characterisation followed the method proposed by GreenDelta [20]. In this methodology, all the input flows are modelled as product flows and expressed in monetary values. The social characteristics are counted as output flows and expressed in the 'activity variable' named 'working hours' in this particular case study [15, 20]. All output flows, quantitative and qualitative, are then scaled to the reference flow, which, in this study, is expressed in Australian dollars. This is achieved using the so-called 'Social Hot Spot Database Workers Hours Model' proposed by GreenDelta [20, 71].

The Social Hot Spot Database Workers Hours Model ranks Country Specific Sectors (CSSs) within supply chains by labour intensity, not monetary value [15]. To estimate the worker hours, this methodology uses two data sources: total wages paid by the CSS per dollar of output and hourly average wages paid by the same CSS to workers [15, 20]. To calculate the worker hours variable, the total wages paid by the CSS are divided by the average wages paid by the hour to workers [15, 20, 71]. Table 6 shows the worker hours of the ACF model and the sectors involved in its production cycle.

Table 6 Social inventory

Sector	Worker hours
Beef feedlot	0.00065
Sorghum production	0.0230
Barley production	0.0110
Water supply	0.0037
Dipping cattle	0.0038
Electricity	0.0051
Diesel	0.0091
Natural gas	0.0006
Vaccination	0.0430
Cattle transportation	0.0024
Mineral supplements	0.0069

2.5 *Life Cycle Sustainability Impact Assessment*

2.5.1 Environmental Impact Assessment

The environmental impact assessment of the ACF model was carried out using the software OpenLCA 1.7.0. beta. The calculations involved in this procedure were performed using the following methods:

- Impact Assessment Method—Chain Management by Life Cycle Assessment (CMLCA) (baseline version 4.4) created by the University of Leiden [4],
- Normalisation and weighting—World, 2000 (year) [4],
- Characterisation factors—CMLCA-IA [21],
- Allocation method—physical allocation,
- Data quality schema—Ecoinvent data quality system [70],
- Aggregation type—weighted average [70] and
- The environmental impact analysis evaluated 11 impact category indicators contained in the CMLCA impact assessment method (Table 7). The baseline version of this impact assessment method contains the impact categories that are most frequently used during LCA analysis [4].

Table 7 LCSA impact assessment categories

Environmental impact category	Cost impact category	Social impact category
Acidification potential (average Europe)	Animal purchase	Child labour
Climate change (GWP100)	Energy	Community engagement
Depletion of abiotic resources (elements, ultimate reserves)	Labour	Equal opportunity
Depletion of abiotic resources (fossil fuels)	Other inputs	Fair salary
Eutrophication (generic)	Transportation	Freedom of association and bargain
Freshwater aquatic ecotoxicity (FAETP inf)	Veterinary products	Gender equality
Human toxicity (HTP inf)	Waste management	Healthy and safety
Marine aquatic ecotoxicity (MAETP inf)	Water	Indigenous rights
Ozone layer depletion (ODP steady state)	Overhead costs	Injuries and fatalities at work
Photochemical oxidation (high Nox)		Local community
Terrestrial ecotoxicity (TETP inf)		Local employment; Working conditions; Working hours

2.5.2 Cost Impact Assessment

To calculate the life cycle costs of the ACF model, a cost model and an impact assessment method were created using OpenLCA 1.7.0. beta and following the models proposed by [32, 52]. The costs categories analysed in this study were selected following the model proposed by [52], which recommends four levels of cost categories when performing LCC: economic costs, life cycle stages, activity types and other cost categories [52]. The impact categories for cost calculation included in the LCC impact assessment method were formulated using the evidence presented in several peer-reviewed studies [52, 85] and are presented in Table 8.

2.5.3 Social Impact Assessment

To measure the social impacts of beef feedlot production in Australia, this study proposed a social impact assessment methodology based on the *Guidelines for Social Life Cycle Assessment of Products* [91] methodology and other methods proposed by [15], Chen and Holden (2016) and [71]. Table 8 shows the impact categories and social/socio-economic subcategories used to perform the SLCA in this study. These categories were selected following the recommendations included in the studies and guidelines reviewed during the elaboration of this research. Based on data availability and the aspects of beef feedlot production in Australia, 13 impact indicators or categories were selected to simulate the social impacts created during the production processes of the ACF model.

To develop the social impact assessment method, the case study also used the OpenLCA 1.7.0. beta software. This impact assessment was named 'Social Impact Assessment AUS' and used the selected impact indicators for the analysis. For each social impact category, a reference unit and impact assessment factors are given. The impact factors vary from low to high risk, where a factor of 0.1–5 is considered low risk, 5–10 is medium risk and above 10 is high risk.

2.6 Sustainability Assessment Results

2.6.1 Environmental Impact Assessment

In this case study, the sustainability of beef production in Australia using a cattle feedlot production system was analysed using the proposed sustainability assessment framework (FSSAF). The environmental and socio-economic impacts associated with every stage of the production of 1 kg of BWG by the ACF model were calculated using an LCSA approach. Table 8 shows the environmental impacts (excluding manure and enteric fermentation emissions) of the sustainability assessment and the results are further discussed in the following sections of this chapter.

Table 8 LCSA environmental impact assessment results

	Reference unit	Results
Environmental impact category		
Acidification potential	kg SO_2 eq	0.0045
Climate change (GWP100)	kg CO_2 eq	2.23
Depletion of abiotic resources (elements, ultimate reserves)	kg antimony eq	0
Depletion of abiotic resources (fossil fuels)	MJ	0
Eutrophication (generic)	kg PO_4-eq	0.61
Freshwater aquatic ecotoxicity (FAETP inf)	kg 1,4-dichlorobenzene eq	0.0018
Human toxicity (HTP inf)	kg 1,4-dichlorobenzene eq	0.0003
Marine aquatic ecotoxicity (MAETP inf)	kg 1,4-dichlorobenzene eq	0.000081
Ozone layer depletion (ODP steady state)	kg CFC-11 eq	0
Photochemical oxidation (high Nox)	kg ethylene eq	0.0000033
Terrestrial ecotoxicity (TETP inf)	kg 1,4-dichlorobenzene eq	0.0017
Costs impact category		
Animal purchase	AUD$	1.4
Energy	AUD$	0.65
Labour	AUD$	0.051
Other Inputs	AUD$	1.84
Transportation	AUD$	0.067
Veterinary products	AUD$	0.033
Waste management	AUD$	0.003
Overhead costs	AUD$	0.52
Water	AUD$	0.105
Social impact category		
Child labour	CH Labour med risk/WH	0.035
Community engagement	CE med risk/WH	0.07
Equal opportunity	EO med risk/WH	3.51
Fair salary	FS med risk/WH	3.48
Freedom of association and bargain	FB med risk/WH	0.10
Gender equality	GE med risk/WH	3.49
Healthy and safety	HS med risk/WH	3.48
Indigenous rights	IR med risk/WH	0.04
Injuries and fatalities at work	IF med risk/WH	3.51
Local community	LC med risk/WH	0.03
Local employment	LE med risk/WH	0.06
Working conditions	WC med risk/WH	1.75
Working hours	WH med risk/WH	3.48

The total GHG emissions generated during the production of 1 kg of BWG by the ACF beef feedlot system was 12.53 kg CO_2 eq/kg BWG including manure and enteric fermentation emissions. These emissions were mainly generated during feed production and machinery operations, and from excreted manure. The production of feed (sorghum silage and barley) accounted for more than 16.8% of these emissions. Manure emissions were calculated using the FarmGAS Calculator ST tool. Methane emissions produced by enteric fermentation are the major source of GHG emissions and also accounted for more than 57% of the total manure GHG emissions (10.3 kg CO_2 eq/kg BWG) generated by the fermentation of fresh manure (5.9 kg CO_2 eq/kg BWG). Nitrous oxide (N_2O) emissions from manure and urine accounted for 37.8% of the total emissions (3.9 kg CO_2 eq/kg BWG).

The acidification of aquatic environments is mainly caused by anthropogenic carbon dioxide and chemicals emitted to the atmosphere that returns to the environment in many different forms [95]. Beef production systems generate large quantities of such substances, such as ammonia, sulphur dioxide, nitrogen oxide and methane [26]. The main sources of atmospheric pollution (Acidification Potential) generated by the ACF model, which, in this study, is expressed in kg of sulphur dioxide equivalents (SO_2 eq), were feed production and machinery operations. Barley production presented the highest potential impact of 0.00071 kg SO_2 eq/kg BWG followed by sorghum silage and tractor operations with 0.00026 kg SO_2 eq/kg BWG and 3.2 E−6 kg SO_2 eq/kg BWG, respectively.

Livestock production contributes to the eutrophication of aquatic systems in Australia [27]. Methane emissions generated by enteric fermentation of intensive animal farming and other activities involved in feedlot operations could increase the occurrence of eutrophication in aquatic systems [27]. In the ACF model, the eutrophication intensity was found to be 0.61 kg PO_4 eq/kg BWG. Again, feed production presented the highest levels of eutrophication potential. Sorghum silage production contributed 87.57% of the total Eutrophication Potential (EP) of the ACF model, followed by barley production, which contributed 10.37%.

The value of the freshwater aquatic ecotoxicity of the ACF model was 0.0018 kg 1,4-dichlorobenzene eq/kg BWG. Of this, 0.00052, 0.00036 and 8.9E−8 kg 1,4-dichlorobenzene eq/kg BWG, were due to sorghum silage, barley production and tractor operations, respectively. With respect to human toxicity, the potential impact of the ACF model was 0.0003 kg 1,4-dichlorobenzene eq/kg BWG. The value of marine aquatic ecotoxicity potential per unit of BWG was 8.1E−5 kg 1,4-dichlorobenzene eq. According to the results, the Terrestrial ecotoxicity of the ACF model was 0.0017 kg 1,4-dichlorobenzene eq/kg BWG. Last, the photochemical oxidation potential of the ACF model was 3.3E−6 kg ethylene eq./kg BWG. The results of the environmental impact assessment of the ACF model are further discussed and compared in the discussion section of this chapter.

2.6.2 Production System Costs

The economic analysis of the ACF system was performed using an LCC method, which has been used to assess the full costing of goods and services and to investigate the costs of products under different scenarios [94]. The life cycle costs of the ACF system are presented in Table 9. According to these results, the total cost to produce 1 kg of BWG was AUD$4.68. Animal purchases to be fattened in the feedlot system represented 29.91% of the total costs. Other inputs (feed costs) contributed 39.81% (AUD$1.84) of the total production cost. Energy and overhead costs (land price, finance costs, capital depreciation, tax, rates and repairs, and maintenance) accounted for the second- and third largest life cycle costs of the system AUD$0.65 (13.88%) and AUD$0.52 (11.11%), respectively.

2.6.3 Social Assessment

The social assessment conducted in this case study attempted to evaluate the social impacts of beef cattle feedlot production in Australia using an SLCA approach. The SLCA method still faces many challenges related to its methodological components and standardisation [60]. However, to perform the social analysis, this case study followed the SLCA guidelines and methodologies proposed by UNEP and several authors who have been working to improve the approach. Therefore, the results of the social analysis should be carefully analysed if intended for use in replication work or policy recommendations.

For this reason, the SLCA results are presented and discussed in this section but the case study is cautious in providing clear deductions concerning the social performance of the beef cattle feedlot industry in Australia. The social impact analysis revealed that certain social impact assessment categories—'Equal opportunity', 'Fair salary', 'Gender equality', 'Health and safety', 'Injuries and fatalities at work' and 'Working hours'—have the highest social risk values of the ACF model, which are expressed in worker hours. However, in accordance with the social impact assessment factors proposed in this current study, they are considered low risk (Table 9).

The other impact categories included in the LCSA social impact analysis also presented lower values; therefore, they are considered low risk in accordance with the social impact assessment methodology used in this case study. Further discussion and comparison of the results are provided in the discussion section of this chapter.

2.7 *Proposed Recommendations and Analysis for Sustainable Beef Production in Feedlots*

In this section, the case study presents and discusses the benefits of the implementation of sustainable beef production processes to increase the sustainability of beef

Table 9 Environmental inventories of the NT8c and SBP models

Flows	Amount/kg BWG	Unit	References
Model NT8c			
Algae (*Scenesdesmus dimorphus* NT8c)	7.28	kg	[81]
Diesel burned in electric generators	0.04	MJ	[38], Wiedmann et al. [105]
Dipping cattle	1	item(s)	[64]
Electricity	5.5	MJ	[38], Wiedmann et al. [105]
Mineral supplements	0.7	kg	[65]
Natural gas QLD	1.05	MJ	[38], Wiedmann et al. [105]
Occupation, agriculture	11	m^2*a	[64, 65, 98]
Solid manure loading and spreading by hydraulic loader and spreader	6.1	kg	[90]
Sorghum silage	5.1	kg	Wiedmann et al. [105]
Tractor engine operations in Australia	0.0095	kg	[90]
Transportation	300	t*km	Davis and Watts [29], [98]
Vaccination	1	item(s)	[25, 80]
Water	73	kg	Davis and Watts [29], [33, 79], Wiedmann et al. [105]
Model SBP			
Algae (*Scenesdesmus dimorphus* NT8c)	7.28	kg	[81]
Anaerobic digestion of manure	0.14	m^3	[46, 86]
Anaerobic digestion plant, agriculture, with methane recovery	1	p	Schleiss and Jungbluth [82]
Biodiesel algae	0.04	kg	[43]
Dipping cattle	1	item(s)	[64]
Electricity production Photovoltaic 570kWp AU	5.5	MJ	Treyer and Vadenbo [89]
Mineral supplements	0.7	kg	[65]
Occupation, agriculture	11	m^2*a	MLA [64, 65, 98]
Sorghum silage	5.1	kg	[105]
Tractor engine operations in Australia	0.0095	kg	[90]

(continued)

Table 9 (continued)

Flows	Amount/kg BWG	Unit	References
Transportation	300	t*km	Davis and Watts [26], [98]
Vaccination	1	Item(s)	[25, 80]
Water	73	kg	Davis and Watts [29], [33, 79], Wiedmann et al. [105, 106]

cattle feedlot systems in Australia as well as increase the material circularity and cascading within the system. The recommended changes to beef production processes are based on the sustainability assessment of the ACF model, the current policies governing beef production in Australia and the increasing emphasis on material circularity within production systems. At this point, it is important to emphasise that to increase the sustainability, circularity and resource-improved efficiency of the beef sector, stakeholder engagement at every stage of the supply chain is essential. The beef production system is complex and integrates several industries, including animal feed production, transportation and veterinary products.

To evaluate the environmental and socio-economic impacts created by the implementation of sustainable beef production technologies in Australia, the case study used a scenario analysis approach. The sustainability assessment framework (FSSAF) proposed by this research project was applied to evaluate and quantify these impacts.

The scenario analysis approach created two different models where sustainable processes, technologies and circularity principles were included in the ACF production system. These models were created using the ACF model parameters and production processes and aimed to represent the operation of the ACF model with sustainable practices integrated into its production system. The following sections of this chapter present and discuss the characteristics and parameters of the ACF model scenarios along with their sustainability analysis using the LCSA approaches included in the FSSAF.

2.7.1 Scenario Descriptions

As previously mentioned, the models represented the ACF model operating using sustainable technologies and circularity principles. The characteristics (including operational and production parameters) of the models are identical; however, the inputs and outputs of the production system differ significantly. With respect to the sustainability analysis, the system boundaries and functional unit remain the same. The LCSA analysis was performed following the procedures specified by the FSSAF methodology for each of the life cycle-based approaches incorporated in the assessment.

In the first model (scenario), the case study built a model called 'ACF algae NT8c as feed (NT8c)'. In this model, barley, which is one of the main feed concentrates supplied to the finishing animals in the ACF model, was replaced by microalgae feed produced by on-farm algal ponds on the feedlot property. Microalgae could become a cheap source of protein to supplement cattle if grown on farm and support producers to become more independent of fluctuations in feed availability and prices [81]. The design and parameters of the on-farm algal ponds to produce the microalgae in this model were based on the system proposed by [81].

The implementation of this model has also other advantages. Once the microalgae feed production system is established, modifications in the system could increase its efficiency, material circularity and decrease the environmental impacts created during beef production as well as generate additional income for feedlot operators [44, 69, 96]. In regards to decreasing environmental impacts, microalgae can be produced using wastewater during feedlot operations, decreasing water requirements during its production processes and environmental impacts created during wastewater management procedures [44, 69]. The cultivation of microalgae also reduces atmospheric emissions through carbon sequestration and could decrease natural waterway pollution and freshwater use once wastewater from different sources can be used in algaculture systems [96]. [69] argued that microalgae production is a sustainable process for the bioremediation of wastewater. Additionally, the author studied the several uses of microalgae such as its ability to remove nitrogen and phosphorus contents in manure and manure effluents, avoiding nutrient pollution in agricultural land and groundwater contamination. Additionally, the surplus of microalgae production in the system, if any, could be used to produce a diversified range of products such as food additives and fertilisers, generating additional income for feedlot operators [69, 96].

The second model was named 'Sustainable beef production in Australian feedlot (SBP)' (Fig. 3). In this model, not only had algal feed replaced barley as one of the main ingredients of the concentrate, but renewable energy and fuel were also introduced into the production system to substitute fossil fuels and coal-fire electricity. The case study assumed that the electricity used to operate the equipment used to supply feed and water to animals, manage waste, perform cleaning and other daily procedures in the SBP model was supplied by a renewable source produced using by-products and resources generated within the feedlot system. A 570 kWp open-ground photovoltaic plant was assumed to produce electricity. The fuel, mainly diesel, necessary to operate equipment and machinery was also assumed of being supplied by a renewable source: biodiesel from algae was presumed to power the equipment and machinery. The biodiesel production processes and their inputs and outputs were based on the studies published by [43, 46]. Lastly, the natural gas (LPG) required to operate equipment and machinery in the ACF model was substituted by biogas produced by anaerobic digestion of the manure generated during the feedlot operations. The case study assumed that an anaerobic digestion plant constructed in the feedlot compound produces the biogas using the manure generated during SBP model operations. The biogas would be produced in a high-rate anaerobic digester system [99].

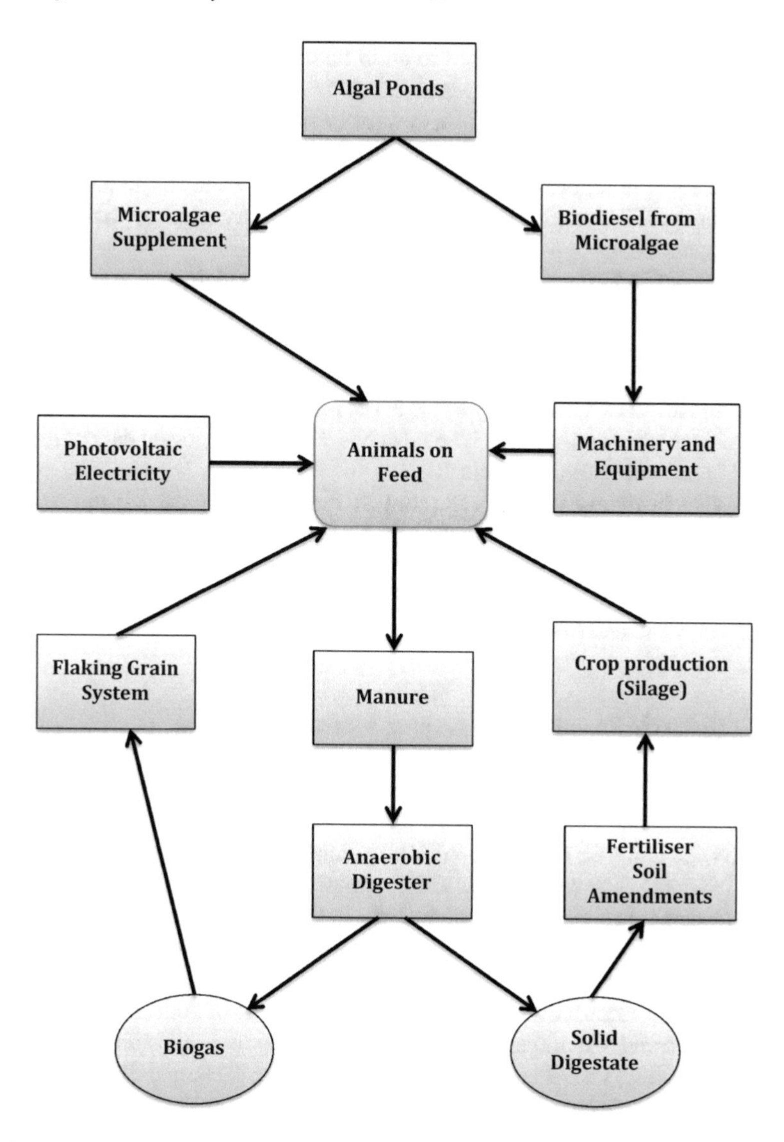

Fig. 3 Production processes of the SBP model

2.7.2 Life Cycle Sustainability Assessment of the NT8c and SBP Models

Life Cycle Sustainability Assessment Inventory

The life cycle sustainability inventory of the proposed models was produced following the principles and approaches of the FSSAF methodology developed in this research. Similar to the method used to build the ACF model environmental inventory, the environmental data to produce the environmental inventories of the NT8c and SBP models were obtained from several peer-reviewed studies and databases as referenced. The process requirements of the inventory were then selected and normalised for a functional unit of 1 kg BWG. Table 9 shows the environmental inventory of the NT8c and SBP models and the input flows required to produce 1 kg BWG in each model.

To create the LCC inventory of the NT8c and SBP models, the case study again followed the LCC method proposed by [52], which was included in the FSSAF methodology. The life cycle costs of the two analysed models were assumed based on data collected and used to produce the LCC inventory of the ACF model. Additional data to produce the LCC inventory of the NT8c and SBP models were collected from published peer-reviewed studies and data cubes containing relevant data of the novel sustainable technologies (see Table 10).

The social inventory of the NT8c and SBP models was carried out using the social analysis methods proposed in the FSSAF methodology. They follow the guidelines and techniques of the model named 'Social Hot Spot Database Workers Hours Model', which was proposed by GreenDelta [20, 71]. Table 11 shows the worker hours of the NT8c and SBP models.

2.7.3 Life Cycle Sustainability Impact Assessment

The environmental impact analysis showed that the NT8c and SBP models have a lower environmental impact than the ACF model. In terms of GHG emissions, the models (NT8c and SBP) cause 10.76% less emissions when compared to the ACF model (Table 12). The reduction of GHG emissions verified in the LSCA was due to the replacement of barley with algae feed in the ration supplied to the finishing animals in the NT8c and SBP models and the replacement of fossil fuels, coal-fired electricity and natural gas by algae biofuel, biogas (anaerobic manure digestion) and photovoltaic electricity in the SBP model. In terms of Acidification Potential (AP), the NT8c and SBP models generate 91.7 and 88% less emissions than the ACF model, respectively. The results also demonstrate that the NT8c and SBP models were considerably better (of lower impact) than the ACF model in all other environmental impact categories analysed by the environmental assessment.

Considering the economic perspective, the SBP model achieved the best performance in most of the analysed life cycle cost impact categories (see Table 13). Consequently, this model has lower life cycle costs throughout its production system due to the savings generated by the on-farm sustainable production processes. The electricity generated by the photovoltaic system has no associated costs and only requires small repairs and maintenance, which can be easily performed by feedlot employees. A similar situation occurs with the anaerobic digester system and the facilities that produce algal feed and biofuel. At this point, it is important to emphasise that the costs of the construction and installation of these systems were not included in the economic analysis performed in this case study.

Table 10 Life cycle costs inventory of the NT8c and SBP models

	Amount (AUD)	References
Model NT8c: flows		
Algae (*Scenesdesmus dimorphus* NT8c) $	9.1	[81]
Animal purchase	1.4	[59]
Capital depreciation	0.019	[7]
Cattle transportation C	0.024	[7, 49, 59]
Diesel burned in electrical generator C	0.0025	[30, 59]
Dipping cattle C	0.01	[49, 59]
Electricity C	0.64	[7, 59]
Finance costs	0.14	[7, 49]
Labour C	0.051	[7, 49, 59]
Land price	1.60E-05	[7, 49, 59]
Manure management	0.003	[30, 59]
Mineral supplements C	0.011	[7, 49, 59]
Rates	0.1	[7, 49, 59]
Repairs and maintenance	0.02	[7, 49, 59]
Sorghum sillage production C	0.37	[49]
Tax	0.25	[7, 49, 59]
Tractor operations C	0.0123	[7, 49, 59]
Tractor operations C	0.031	[7, 49, 59]
Vaccination C	0.012	[59]
Water use C	0.105	[59]
Model SBP		
Model SBP: flows		
Algae (*Scenesdesmus dimorphus* NT8c) $	9.1	[81]
anaerobic digestion of manure $	0	[86]
Animal purchase	1.4	[59]
Biodiesel algae $	0	[43, 46]
Capital depreciation	0.019	[30, 59]
Cattle transportation C	0.024	[49, 59]
Dipping cattle C	0.01	[7, 59]
Electricity production Photovoltaic 570 kWp AU$	0	[89]
Finance costs	0.14	[7, 49, 59]
Labour C	0.051	[7, 49, 59]
Land price	1.60E−05	[30, 59]
Manure management	0.003	[7, 49, 59]

(continued)

Table 10 (continued)

	Amount (AUD)	References
Mineral supplements C	0.011	[7, 49, 59]
Photovoltaic system 570 kWp $	0	[89]
Rates	0.1	[7, 49, 59]
Repairs and maintenance	0.02	[7, 49, 59]
Sorghum sillage production C	0.37	[49]
Tax	0.25	[7, 49, 59]
Tractor operations C	0.0123	[7, 49, 59]
Tractor operations C	0.031	[7, 49, 59]
Vaccination C	0.012	[59]
Water use C	0.105	[59]

Table 11 Social inventory of the NT8c and SBP models

Sector	ACF Worker hours	NT8c worker hours	SBP worker hours
Beef feedlot	0.00065	0.00065	0.00065
Sorghum production	0.023	0.023	0.023
Barley production	0.011	0	0
Water supply	0.0037	0.0037	0.0037
Dipping cattle	0.0038	0.0038	0.0038
Electricity	0.0051	0.0051	0
Diesel	0.0091	0.0091	0
Natural gas	0.0006	0.0006	0
Vaccination	0.043	0.0043	0.0043
Cattle transportation	0.0024	0.0024	0.0024
Mineral supplements	0.0069	0.0069	0.0069
Algae (*Scenesdesmus dimorphus* NT8c)	0	0.29	0.29
Electricity production Photovoltaic 570 kWp	0	0	0.2
Biogas (anaerobic digestion of manure)	0	0	0.2
Biodiesel (Algae biofuel)	0	0	3.14

The social impact assessment results of the NT8c and SBP models are presented in Table 14. Again, the SBP model achieved the best performance. According to the social impact assessment, the SBP model causes less social impacts in categories evaluated in the SLCA. Although this research attempted to construct a sound social impact assessment methodology to evaluate the social impacts of products

Table 12 Environmental impact assessment results of the NT8c and SBP models

Impact category	Reference unit	ACF model	NT8c model	SBP model
Acidification potential (average Europe)	kg SO_2 eq	0.0045	0.00037	0.00054
Climate change (GWP100)	kg CO_2 eq	2.23	1.99	1.99
Depletion of abiotic resources (elements, ultimate reserves)	kg antimony eq	0	0	0
Depletion of abiotic resources (fossil fuels)	MJ	0	0	0
Eutrophication (generic)	kg PO_4 eq	0.61	0.55	0.54
Freshwater aquatic ecotoxicity (FAETP inf)	kg 1,4-dichlorobenzene eq	0.0018	0.0014	0.00052
Human toxicity (HTP inf)	kg 1,4-dichlorobenzene eq	0.0003	0.000021	0.0000015
Marine aquatic ecotoxicity (MAETP inf)	kg 1,4-dichlorobenzene eq	0.000081	0.000066	0.00000013
Ozone layer depletion (ODP steady state)	kg CFC-11 eq	0	0	0
Photochemical oxidation (high Nox)	kg ethylene eq	3.3E-06	3.60E-06	6.70E-07
Terrestrial ecotoxicity (TETP inf)	kg 1,4-dichlorobenzene eq	0.0017	0.0012	0.000078

Table 13 Life cycle costs impact assessment results of the NT8c and SBP models

Costs impact category	Reference unit	ACF model	NT8c model	SBP model
Animal purchase	AUD	1.4	1.4	1.4
Energy	AUD	0.65	0.65	0
Labour	AUD	0.051	0.051	0.051
Other inputs	AUD	1.84	0.37	0.37
Transportation	AUD	0.067	0.067	0.529016
Veterinary products	AUD	0.033	0.033	0.0673
Waste management	AUD	0.003	0.003	0.033
Overhead costs	AUD	0.52	0.52	0.003
Water	AUD	0.105	0.105	0.105

Table 14 Social impact assessment results of the NT8c and SBP models

Impact category	Reference unit	ACF model	NT8c model	SBP model
Child labour	CH labour med risk	0.035141169	0.001331169	0.001002494
Community engagement	CE med risk	0.071299406	0.037489406	0.004621906
Equal opportunity	EO med risk	3.516475604	0.135475604	0.102608104
Fair salary	FS med risk	3.485452394	0.104452394	0.088130119
Freedom of association and bargain	FB med risk	0.101461884	0.101461884	0.085141884
Gender equality	GE med risk	3.496106909	0.115106909	0.082239409
Healthy and safety	HS med risk	3.482763867	0.101763867	0.101435192
Indigenous rights	IR med risk	0.043763256	0.009953256	0.009399356
Injuries and fatalities at work	IF med risk	3.512145009	0.131145009	0.098277509
Local community	LC med risk	0.035281456	0.001471456	0.001031306
Local employment	LE med risk	0.063920109	0.030110109	-0.002532166
Working conditions	WC med risk	1.757367322	0.066867322	0.050545047
Working hours	WH med risk	3.480709899	0.099709899	0.083387624

and production systems, there are issues in the application of SLCA methodologies, not just in this particular case study but also in work by other researchers and life cycle analysis practitioners. These issues were identified and analysed during the literature review performed to design the FSSAF methodology and have been previously mentioned in this chapter. To validate and support the social impact assessment results, a literature review assessed possible social issues caused by the feedlot industry in Australia. Additionally, the review investigated the possible effects of the implementation of anaerobic manure digestion and the production of algae feed and biodiesel in the feedlot industry. The results of the review are presented and discussed in the following section.

3 Discussions

The cattle feedlot sector in Australia supplies both domestic and international markets and plays an important role in the Australian economy. Approximately, 95% of Australian feedlots are family owned and they has a production value of more than AUD$4.6 billion (ALFA 2020). The feedlot industry directly and indirectly employs approximately 28,500 workers [10]. In Australia, cattle feedlot systems are mostly located in the state of Queensland (60%), however, this type of beef production system is also present in other states [10]. Beef feedlot systems are diverse and their production processes vary to different degrees depending on the system selected by the management team [64, 105]. Due to the diversity of cattle feedlot production systems, the lack of similar studies and the different environmental and socio-economic analysis methodologies adopted in these studies (particularly LCA methodologies), it is difficult to compare the results of this case study with those of other peer-reviewed studies. However, for the sake of accuracy, quality and support of the sustainability assessment results of the ACF model, the case study compared the ACF model results against other available data where possible.

Evaluation of the sustainability of cattle feedlot systems is fundamental for predicting environmental issues and implementing measures to prevent problems of similar nature in the future. Prior to this current study, a limited number of LCA studies were performed to identify the environmental impacts created by cattle feedlot systems in Australia. These studies were conducted by [76, 103, 106] and [105]. The authors mainly used the LCA approach to calculate resource use and GHG emissions and did not include other sustainability impact category indicators in their research. To assess the other environmental impact results of the ACF model, the case study compared them with results obtained by peer-reviewed studies that analysed the environmental impacts of cattle feedlot production in other countries (Table 15). However, it is important to emphasise once again that LCA results are difficult to be compared due to several reasons previously mentioned in this chapter.

The main source of GHG emissions from beef production systems is methane from enteric fermentation, manure management procedures [76, 102]. Emissions of GHG also arise in beef systems during fertilisation procedures applied in feed production and fossil fuel use for transportation and energy production [53, 102]. According to Wiedemman et al. 102, enteric methane is the largest contributor to GHG emissions of the Australian agricultural sector. The authors argued that methane contributes to more than 67% of the national emissions. Manure management activities are the second major contributor to emission intensity at the farm level in beef production systems. Methane and nitrous oxide (N_2O) are the main source of GHG emitted during manure management procedures [83, 101]. Additionally, large quantities of GHG are emitted during fertilisation activities in areas used to produce feed rations supplied to finishing animals. Feed production in Australia uses considerable amounts of nitrogen-based fertilisers to increase crop production [76]. Nitrogen-based fertilisers release N_2O, which depletes the ozone layer and increases the effects of global warming (Bell et al. 2011; [83]). Lastly, another major source of GHG emissions from beef production results from energy consumption required during production processes (electricity, natural gas and fuel use) and transportation of animals and animal feed [53, 75].

As projected, and in line with the other LCA studies of beef production evaluated in this study, the main source of GHG in the ACF model is generated by the animals on feed and feedlot feed production. Enteric fermentation was responsible for more than 47% of the total GHG emissions of the ACF model, while the studies conducted by Wiedemman et al. [101, 105] enteric fermentation accounted for 70% and 50% of the total GHG emissions generated in the systems. The study conducted by [55] also verified that enteric emissions contributed to more than 65% of the GHG emissions associated with the beef production system evaluated in their research.

Table 15 Comparison of environmental impacts of LCA research on beef feedlot in Australia and worldwide

Impact category indicator	Asem-Hiablie et al. [11]	[72]	[55]	[76]	[106]	[103]	[105]	ACF model
Acidification potential (kg SO_2 eq.)	0.726	N/A[a]	0.328	N/A	N/A	N/A	N/A	0.0045
Climate change (GWP)	48.4 kg CO_2 e/kg CB[b]	32.3 kg CO_2 e/kg LWG	23 kg CO_2 e/kg CW[c]	5.5 CO_2 e/kg HSCW[d]	7.5 kg CO_2 e/kg LWG	23.7 kg CO_2 e/kg boneless beef	4.6 kg CO_2 e/kg LWG	12.53 CO_2 e/kg BWG
Depletion of abiotic resources (elements, ultimate reserves kg antimony eq.)	1.03E−4	N/A	N/A	N/A	N/A	N/A	N/A	0
Depletion of abiotic resources (fossil fuels; MJ)	N/A	N/A	N/A	N/A	N/A	N/A	N/A	0
Eutrophication (generic; kg PO_4-eq.)	N/A	0.602	0.409	N/A	N/A	N/A	N/A	0.61
Freshwater aquatic ecotoxicity (FAETP inf)	N/A	N/A	N/A	N/A	N/A	N/A	N/A	0.0018

(continued)

Table 15 (continued)

Impact category indicator	Asem-Hiablie et al. [11]	[72]	[55]	[76]	[106]	[103]	[105]	ACF model
Human toxicity (HTP inf)	0.027	N/A	N/A	N/A	N/A	N/A	N/A	0.0003
Marine aquatic ecotoxicity (MAETP inf)	N/A	N/A	N/A	N/A	N/A	N/A	N/A	0.000081
Ozone layer depletion (CFC-11 eq.)	1,686E−6	N/A	N/A	N/A	N/A	N/A	N/A	0
Photochemical oxidation (kg ethylene eq.)	0.018	N/A	N/A	N/A	N/A	N/A	N/A	0.0000033
Terrestrial ecotoxicity (TETP inf)	N/A	N/A	N/A	N/A	N/A	N/A	N/A	0.0017

[a] N/A = Not applicable
[b] CB = Consumer Benefit (1 kg of consumed, boneless edible beef)
[c] CW = Carcass Weigh
[d] HSCW = Hot Standard Carcass Weight

With respect to total GHG emissions, including manure emissions, this case study presented similar results found by [76, 106] and Wiedemman et al. [105] (5.5 kg CO_2 e/kg Hot Standard Carcass Weight, 7.5 kg CO_2 e/Kg Live Weight Gain and 4.6 CO_2 e/kg LWG, respectively). However, the ACF emissions were much lower than the 23.7 kg CO_2 e/kg boneless beef reported by [11, 72, 103] and [55] (48.4 kg CO_2 e/kg CB, 32.3 kg CO_2 e/kg LWG and 23 CO_2 e/kg carcass weight, respectively). We believe the variations presented in the GHG emission results when compared to other studies examined in this research are also due to other reasons. Emission intensity fundamentally varies with differences in beef production processes affecting the system emissions [105]. Those factors directly affect the GHG contribution analysis increasing or decreasing the impacts associated with feed production, transportation, enteric fermentation and manure management. Another reason for variations in the results is the emission estimation method used in the environmental assessment (Wiedemman et al. 2016, [62]). The prediction of emission intensity in a particular system is highly sensitive to the environmental impact assessment method utilised to perform the sustainability assessment or environmental impact assessment of the system. Wiedemman et al. [105] verified that the method used to calculate enteric fermentation emissions in the Australian GHG inventory is more conservative and predicts higher emissions than the IPCC method, which is the international default.

The study conducted by Wiedemman et al. [101] analyses all the stages of the Australian red meat export supply chains to the USA. The differences in the results when compared to the ACF model are likely due to the different feed rations, production systems and transportation distance or 'food miles' applied in the two studies. According to the results, meat processing (4%) was the second largest contributor of GHG in the systems followed by transportation of meat products from Australia to distribution warehouses in the USA (3%) [101]. In regards to the study performed by Asem-Hiablie et al. [11], the disparity in the results when compared to the ACF model is mainly governed by differences in the beef production system evaluated in the study, the impact assessment method and system boundaries. The authors investigated the environmental impacts of the entire beef supply chain in the USA including the life cycle of a cow-calf and feedlot system from birth to consumers using the environmental impact metrics included in the BASF LCA methodology [14]. However, the GHG intensity of the entire supply chain is considerably higher than the ACF model; the beef feedlot production phase of the system (feed production and finishing) presented similar results (13.63 CO_2 e/kg CB) [11].

With respect to AP, it is important to quantify its impacts in beef production systems once manure management and fertilisation procedures of animal feed production release gases that cause acidifying effects to the environment (water and soil acidification and acid rain) [78]. Manure storage produces ammonia (NH_3) emissions, and the spreading of manure in fields produces N_2O emissions through volatilisation. Fertilisation procedures using nitrogen (N) fertilisers applied in crop production destined to animal feed directly contribute to N_2O emissions from agricultural soils [19, 78]. AP is associated with the deposition of acidifying contaminants in aquatic and terrestrial ecosystems [95]. Acidification, particularly of aquatic environments, is an environmental impact that threatens the functionality of aquatic and terrestrial ecosystems [95]. Acidification of aquatic environments occurs when gaseous substances such as sulphur dioxide (SO_2), nitrogen oxides (NOx), nitrogen monoxide (NO), nitrogen dioxide (N_2O) and other various substances emitted to the atmosphere mix and react with water, other gases and chemicals forming acidic pollutants that return to the environment in many different forms [19, 78].

The AP result analysis of the ACF model presented lower values than those obtained by [55] and Asem-Hiablie et al. [11] in finishing cattle systems in the United States (US). The analysis of the results concluded that this difference was due to the feedlots evaluated in these studies having different production and management systems than those applied in the ACF model. [55] analysed a beef cattle feedlot production system in the Northern Great Plains of the US. Their LCA analysis included the upstream inputs and processes involved in the management of cattle breeding before entering the feedlot. These processes were excluded from this case study. Additionally, the concentrate (feed) supplied to the animals during their finishing period differed from that in the ACF model. The other study [11] used an LCA approach to analyse the environmental impacts of a beef system in the US. In their analysis, the authors adopted an unusual functional unit (Consumer Benefit) for beef LCA studies and included the entire beef supply chain.

The sustainability assessment carried out in this study also measured the EP of the ACF model. Eutrophication Potential (EP) is defined as the potential of excessive nutrient enrichment of water and soil. This metric covers the impacts on terrestrial and aquatic environments due to over-fertilisation or excess amount of nutrients, especially N and phosphorus (P) [97]. The over-supplementation of nutrients rises the growth of plankton algae and other aquatic plants, which increases the consumption of oxygen by bacteriological degradation of dead biomass in aquatic environments, changing the composition of species. The impacts of eutrophication on the terrestrial ecosystem can change the function and diversity of species. According to [97], the reaction of the aquatic ecosystems to the addition of nitrates and phosphate chemical and natural-based substances through agricultural runoff and sewage is one of the main causes of eutrophication.

Eutrophication mainly caused by runoff of nutrients from agricultural activities costs more than 200 million dollars to the Australian economy every year, particularly in regions with high rainfall rates and vast concentration of irrigated systems [27]. Beef production systems also contribute to the eutrophication of the aquatic system in Australia. Nutrient enrichment of waterways caused by phosphorus and nitrogen runoff and leaching from agricultural areas in the country is a common issue [27]. One of the main causes of this environmental problem is the excessive use of phosphorus- and nitrogen-based fertilisers in pastures and agricultural areas which is a common procedure used in Australia (Watkins and Nash 2010). Additionally, methane emissions, largely generated by enteric fermentation and manure management in livestock's grazing and feedlots production system, also increase the occurrence of eutrophication in aquatic systems [27].

With regard to EP expressed in kg PO_4 eq., this study produced similar values to those obtained by [72], which adopted similar system boundaries, although the concentrate supplied to the animals differed significantly. This case study found higher values for EP than those obtained by [55]. In their work, the authors included the breeding herd in the LCA analysis, which is likely the main reason for the differences in the EP values. Moreover, EP depends on the characteristics of the product system analysed and other conditions such as temperature, rainfall and soil structure [31].

The LCSA framework proposed in this study used an LCC approach to analyse the production system costs of the ACF model. Although LCC has been widely used, the technique is still not standardised and clear indicators are not yet described to fully represent a cause-and-effect chain in LCC studies [85]. However, Klöpffer and Ciroth (2011) have stated that real money flows and LCC methods must be incorporated in sustainability analysis of products and production systems.

Based on these arguments, the interpretation of LCC results remains challenging, as does the evaluation of economic impacts during sustainability assessments. This case study presumed that the economic assessment results of the ACF model could be useful during decision-making processes in the management of beef production costs. Moreover, the results of the economic analysis may serve as a benchmark for future studies into the implementation of sustainable technologies for cattle feedlot production in Australia and worldwide.

This case study also attempted to evaluate the social impacts of the ACF model using an SLCA approach. Since few studies have attempted to analyse the social impacts of beef production in general using an SLCA approach, many of the methodological challenges and issues remain. Hence, the social impact results of the ACF model should be interpreted cautiously and the social assessment cannot be considered complete and robust. However, to increase the integrity of the social assessment results, the case study compared them to average Australian socio-economic indicators published by Australian Government statistical agencies (Australian Bureau of Statistics, Safe Work Australia, Australian Institute of Health and Welfare, Fair Work Commission Australia and Torrens University Australia). All social impact assessment indicators analysed were considered low risk. Nevertheless, six indicators (Equal opportunity, Fair salary, Gender equality, Health and safety, Injuries and fatalities at work and Working hours) presented higher values. When compared to Australian statistics, the ACF social assessment results revealed a number of interesting facts. For example, the ACF child labour results were similar to national statistics. According to the ABS, there is no evidence of child employment in Australia. Additionally, the case study analysed the sectors that supply goods and services to the ACF model and the Australian cattle feedlot industry. The results revealed that these sectors are based in Australia and they do not employ under-age workers.

With respect to Equal opportunity, all employers in Australia are legally obligated to prevent workplace discrimination and harassment in accordance with the Equal opportunity Act 2010. Nevertheless, the Committee for Economic Development of Australia (CEDA) recently published a report stating that in 2017, the national gender pay gap in Australia was 15.3%. The social inventory produced in this case study showed that the gender gap pay in the Australian agricultural industry is above the national average: it reached 18.9% during the 2015–16 financial year [100]. The CEDA study also analysed Gender equality in Australia and found that of 19 industry sectors in the country, seven have workforces with less than 30% female employees [18]. Moreover, income inequality is an issue in some sectors. Regarding managerial positions and Working hours (full-time positions), the percentage of females is significantly lower than that of males. The present social inventory results verified that in the Australian livestock industry, female employment comprises 38.5% of the total workforce (ABS 2013).

In relation to Fair salary, the social inventory found that the average weekly earnings for full-time livestock workers is well below the national average (AUD$ 19.49) [2, 3, 77]. Additionally, the unemployment rate is considered high, when compared to the national rates [1]. With respect to Health and safety and Injuries and fatalities at work, the industries involved in agriculture have the highest rates of serious claims per hour worked and the second highest number of work-related fatalities in 2016 [84].

Regarding the scenario analysis, the sustainability assessment results identified areas where changes in the feedlot production system could improve the sustainability of Australian beef production. Implementation of the sustainable approaches proposed in this case study could increase the circularity of materials and waste within the feedlot system and lessen the environmental and socio-economic impacts of its production life cycle. Furthermore, the introduction of sustainable beef production technologies may support the sustainable development of the beef industry and its supply chain, consequently improving the socio-economic conditions of local communities.

Population growth is increasing meat consumption and forcing the livestock industry to expand production to meet this demand. Consequently, the agricultural industry will need to increase its efficiency to supply both food for human consumption and feed for animals [56]. Microalgae have been studied as a sustainable animal feed substitute to decrease the environmental footprint of cattle production systems. According to [56], microalgae are a rich source of the nutrients found in stockfeed supplements and some species have high nutritional value. [23] tested various algae species in comparison to supplements commonly used as feedstuffs for cattle. The latter authors concluded that some micro-algae species (i.e., *Spirulina platensis* and *Chlorella pyrenoidosa*) may effectively be used as feed for ruminant animals.

New, low-cost technology for microalgae production has been developed in Australia to increase production feasibility. [81] developed a novel system of on-property microalgae production; the primary results are promising and several large-scale on-farm projects have been proposed in Australia [81]. This production system was included in the NT8c and SBP models as an algae production technology to replace grains with algae in the supplement offered to finishing animals. The sustainability assessment results showed that the replacement of barley with algae significantly decreases all the environmental impacts of feedlot systems. As previously mentioned, this change would decrease more than 10% of the GHG emissions compared to the conventional system model (ACF model). This change would also lessen the indirect environmental impacts of beef production in feedlots; if algal feed is produced on farm, the environmental impacts of transporting grains for feed supplements will decrease. Additionally, animal feed production competes with crops (for human consumption) for water and land [69]. The replacement of grains by algae in livestock production could support the balance between food, animal feed and biofuel production and decrease natural resource use and input production and consumption during grain production [56, 37]. Furthermore, Costa et al. [22] indicated that microalgae species used as feedstuffs for cattle led to fatty acids formation in the rumen that if transferred to meat could have health-related benefits to consumers.

Lastly, microalgae production might decrease the environmental impacts and costs of waste management in feedlot systems and increase the circularity of materials within the system [44, 67, 69]. Microalgae can be cultivated in agricultural and manure wastewater [44, 67] those microalgae produced in wastewater are known as biofilters and have several economic uses, including fertiliser, animal feed and biofuel production. [46] proposed closed-loop nutrient recycling that integrates biodiesel and biogas production from microalgae. The authors presented different types of microalgae production and treatment to produce a stable oil product for use as biofuel and anaerobic digestion of wastewater to produce biogas (methane). [99] also studied the feasibility and efficiency of several types of anaerobic digestion to produce biogas from solid and liquid manure.

The production and use of algae as feedstuff have social and economic advantages, as shown by the sustainability analysis in this case study and by other researchers and stakeholders in livestock production. The cost analysis of the NT8c and SBP models found that the life cycle cost of 1 kg BWG produced in the simulated feedlot systems decreased by more than 31% and 45% for NT8c and SBP, respectively. This change is due to the substitution of barley in both cases and the implementation of sustainable technologies in the second case. The SBP model has no production costs related to energy use because it uses renewable energy (biogas, solar electricity and algal biodiesel) to power generators and machinery.

According to the literature review performed in this case study, the implementation of sustainable biofuel and energy production technologies in beef feedlot production systems also has numerous socio-economic benefits. The products of these systems could reduce waste management costs, create jobs in remote communities and decrease environmental impacts that affect human health. Algal culture in wastewater could become an important element in water recycling as a biofilter and the algae used to clean the water can be sold as stockfeed and other by-products. Additionally, algae absorb considerable nutrients from wastewater and can be used as fertiliser. Since fertilisers are becoming scarce and expensive, algal fertilisers could become a profitable industry that generates jobs in the near future [13].

In terms of sustainable biofuel production, microalgae biomass can be used for biodiesel production from the high lipid content of various algae species. [50] highlighted that algal biodiesel has several socio-economic advantages when compared to other types of biofuels. Algal biomass grows rapidly, produces more biomass per hectare than plant-based biofuels, is less toxic and does not compete with food production [50]. [107] studied the socio-economic impacts of algal biodiesel development in China and found that the industry's development had multiple socio-economic benefits in terms of both economic and employment growth. Similarly, [34] found considerable benefits from the implementation of algae-based biofuel production in regional areas. Thus, the development of this industry could promote sustainable development in rural communities through job creation and increases in household income, social well-being, and food and energy security [34, 107]. Cost-effective biofuel production might attract investors to regional hubs, creating sustainable economic development and jobs. In terms of energy security, the development of the algal biofuel industry diversifies the energy supply chain and decreases the risk of energy shortages, price fluctuations and dependence on fossil fuels [34].

Biogas production from animal manure is another sustainable technology proposed and analysed in this case study. The implementation of this technology in the beef sector creates beneficial environmental, social and economic impacts, as verified by the sustainability assessment. The results revealed that the replacement of fossil-based gas with manure-based biogas reduced GHG emissions and other environmental impacts of beef production in feedlot systems and also decreased both life cycle costs and social impacts.

Research suggests that the implementation of biogas production from agricultural waste promotes sustainable development and creates socio-economic and health benefits in rural communities [24, 108, 109]. For example, [109] modelled and evaluated a circular manure-based biogas supply chain. In their circular economy model, manure is collected from farms and processed at biogas plants. The biogas produced is used to generate bioenergy (heat and electricity) to supply local markets, while the digestate of the anaerobic digestion process returns to farmers for use as fertiliser to cultivate crops. The implementation of this model could create a closed-loop supply chain and offer environmental and socio-economic benefits to rural communities, small businesses and local governments. [54] analysed the socio-economic benefits of biogas production and utilisation in Central and Eastern Europe. They found that the production of biogas from agricultural waste in the region creates the following environmental, social and economic benefits for the local society:

- Sustainably supplies heat and electricity to local households and farmers;
- Replaces fossil fuels with renewable fuels and therefore, increases the sustainability of the energy supply and prevents energy shortages;
- Protects natural resources and the environment by reducing dependence on fossil fuels, GHG emissions, waste, and soil and groundwater contamination;
- Creates jobs and increases the farmers' income;
- Reduces energy costs;
- Creates a closed-loop waste and nutrient recycling system.

Lastly, the implementation of photovoltaic systems to produce renewable energy to power electric components and equipment in feedlot production systems will certainly create environmental and socio-economic benefits for the industry and its stakeholders. Solar energy production reduces GHG emissions and the electromagnetic radiation produced when energy is transported across electricity grids. Furthermore, the installation and maintenance of photovoltaic systems creates jobs and supports local businesses. Thus, solar systems could reduce beef production costs and prevent energy shortages in remote farms and communities [68].

4 Conclusions and Further Studies

4.1 Conclusions

The results of the case study were used to test the FSSAF framework, which aims to evaluate the sustainability of food systems in Australia and other regions. The case study aimed to use the FSSAF to holistically assess the environmental, economic and social impacts and benefits of the beef industry, focusing on beef produced in feedlot systems. Additionally, it used the FSSAF to model and evaluate scenarios where sustainable technologies were used to increase the sustainability of beef feedlot production in Australia.

The results verified important factors in the sustainability of beef production using feedlot systems in Australian conditions. It revealed the environmental and socio-economic impacts created during the life cycle of beef production and proposed changes to the production structure that improved the predicted sustainability and increased the material circularity of the system. Additionally, to verify the accuracy and robustness of the results, the case study compared its results against those of other studies and reviewed the literature. This analysis also complemented the sustainability assessment of beef production and consideration of sustainable technologies that could increase the socio-economic benefits of beef production supply chains.

Lastly, the results acquired in the case study demonstrated that the use of microalgae in feedlot systems has several benefits. When microalgae is used as feedstuff, it could reduce the environmental impacts created during beef production in feedlot systems and increase the system sustainability by creating jobs and reducing waste management costs. Additionally, microalgae can be used within the system to produce biofuel to fuel machinery and other equipment used during the operation of the feedlot and its facilities. Therefore, according to the results of the case study, the production of microalgae and its use (feedstuff and biofuel) in feedlot production systems proved to be the most beneficial amongst the proposed sustainable technologies and approaches presented in this chapter.

4.2 Further Studies

The agri-food supply chain is one of the most important industries in Australia in terms of economic revenue. Sustainable and efficient production systems are fundamental to maintain the industry's competitiveness, decrease its use of natural resources and reduce the environmental burdens of its production systems. The sustainability assessment framework proposed in this research could be used as a guideline in assessing the sustainability aspects of the Australian beef and food industries. The framework can be combined with other methodologies such as the Australian Beef Sustainability Framework [2] to identify current inefficient processes using an integrated approach or be used to evaluate new technologies to improve the sustainability of food and beef production systems in Australia. The proposed framework could also be altered and integrated with other methods to improve its functionality and flexibility. For example, there is currently no standardised framework to analyse the sustainability of adaptive complex systems; the FSSAF can be integrated with other approaches to evaluate the eco-efficiency, sustainability and competitiveness of food and other production systems. Moreover, it could be useful to construct a comprehensive database containing data related to environmental impacts and costs, resource efficiency, and the economic and social importance of the Australian food industry.

Further studies to enhance the efficacy of the FSSAF could support the dissemination and application of CE principles in the food system. Based on the case study results, the framework proved to be efficient in evaluating the three pillars of sustainability and assess the effects of the implementation of sustainable production processes using the principles of material circularity. Therefore, supplementary research could improve the efficiency of the FSSAF in predicting the impacts of the CE principle inclusion in policy design and implementation providing guidance to policy-makers and stakeholders involved in food production. Additionally, future works to continue the development of the sustainability assessment framework (FSSAF) and the search for sustainable production processes based on the principles of CE could support in solving some issues and challenges pointed out by [45]. The authors argued that the main issues of the implementation of technologies to convert agricultural wastes into feasible bioproducts are to find efficient methodological approaches to holistically analyse the impacts of the implementation of the processes and to predict the benefits of this practice.

References

1. ABS (2017) Labour force, Australia, Australian Bureau of Statistics, Canberra
2. ABSF (2020) Australian beef sustainbility—annual update, Australian Beef Sustainability Framework, Melbourne
3. ACCC (2016) Cattle and beef market study. Australian Competition & Consumer Commission, Canberra
4. Acero AP, Rodríguez C, Ciroth A (2016) LCIA methods: impact assessment methods in Life Cycle Assessment and their impact categories. GreenDelta, Berlin
5. AEMO (2018) Australian Energy Market Operator2018. https://www.aemo.com.au/
7. AFI (2013) Farmgas scenario tool. Australian Farm Institute, Sydney
8. AFI (2014) FarmGas financial tool: user guide. Australian Farm Institute, Sydney
9. Akhtar S, Reza B, Hewage K, Shahriar A, Zargar A, Sadiq R (2015) Life cycle sustainability assessment (LCSA) for selection of sewer pipe materials. Clean Technol Environ Policy 17:973–992
10. ALFA (2018) About the Australian Feedlot Industry, Australian Lot Feeders Association, viewed 20 October 2020. http://www.feedlots.com.au/industry/feedlot-industry/about.
11. Asem-Hiablie S, Battagliese T, Stackhouse-Lawson KR, Alan Rotz C (2018) A life cycle assessment of the environmental impacts of a beef system in the USA. Int J Life Cycle Assess

12. ATTRA (1999) Sustainable beef production. National Center for Appropriate Technology, Fayetteville
13. Australia21 (2017) Opportunitties for an algal industry in Australia, Australia21 Ltd, Weston
14. BASF (2013) Submission for verification of eco-efficiency analysis under NSF protocol P352, Part A. U.S. beef—Phase 1 eco-efficiency analysis, BASF Corporation, Florham Park
15. Benoit-Norris C, Cavan DA, Norris G (2012) identifying social impacts in product supply chains: overview and application of the social hotspot database. Sustainability 4:1946–1965
16. Bond R, Curran J, Kirkpatrick C, Lee N, Francis P (2001) Integrated impact assessment for sustainable development: a case study approach. World Dev 29(6):1011–1024
17. Cabon, DH, Terwijin, MJ, Williams, AAJ (2017) Impacts and adaptation strategies for a variable and changing climate in the CENTRAL QUEENSLAND REGION. International Centre for Applied Climate Sciences, Toowoomba
18. CEDA (2018) How unequal? Insights on inequality, Committee for Economic Development of Australia, Melbourne
19. Chai R, Ye X, Ma C, Wang Q, Tu R, Zhang L, Gao H (2019) Greenhouse gas emissions from synthetic nitrogen manufacture and fertilization for main upland crops in China. Carbon Balance Manage 14(1):20
20. Ciroth A, Duyan Ö (2013) Social hot spots database in open LCA. GreenDelta, Berlin
21. CMLDIE (2016) *CML-IA Characterisation Factors*, University of Leiden Department of Industrial Ecology, https://www.universiteitleiden.nl/en/research/research-output/science/cml-ia-characterisation-factors
22. Costa DFA, Quigley SP, Isherwood P, McLennan SR, Sun XQ, Gibbs SJ, Poppi DP (2020) Chlorella pyrenoidosa supplementation increased the concentration of unsaturated fatty acids in the rumen fluid of cattle fed a lowquality tropical forage. Revista Brasileira de Zootecnia 49
23. Costa DFA, Quigley SP, Isherwood P, McLennan SR, Poppi DP (2016) Supplementation of cattle fed tropical grasses with microalgae increases microbial protein production and average daily gain. J Anim Sci 94(5):2047–2058
24. Cucui G, Ionescu CA, Goldbach IR, Coman MD, Marin ELM (2018) Quantifying the Economic Effects of Biogas Installations for OrganicWaste from Agro-Industrial Sector. Sustainability 10
25. DAF (2016) Livestock vaccination, Department of Agriculture and Fisheries Queensland Goverment, viewed 20 June 2019, https://www.daf.qld.gov.au/business-priorities/biosecurity/animal-biosecurity-welfare/animal-health-pests-diseases/protect-your-animals/livestock-vaccination
26. Davis RJ, Watts PJ (2011b) Environmental sustainability assessment of the Australian feedlot industry. Part A Report: Water Usage at Australian Feedlots, Meat & Livestock Australia Limited, Sydney
27. Davis JR, Koop K (2006) Eutrophication in Australian rivers, reservoirs and estuaries—a southern hemisphere perspective on the science and its implications. Hydrobiologia 559:23–76
28. Davis RJ, Watts PJ, McGahan EJ (2012) Quantification of feedlot manure output for beef-bal model upgrade. Rural Industries Research and Development Corporation, Canberra
29. Davis RJ, Watts PJ (2011a) Environmental sustainability assessment of the australian feedlot industry. Part B—Report energy usage and greenhouse gas emission estimation at Australian Feedlots, Meat & Livestock Australia Limited, Sydney
30. Deblitz C, Dhuyvetter K, Davies L (2012) Benchmarking Australian and US Feedlots. Meat & Livestock Australia, Sydney
31. Dick M, Silva MA, Dewes H (2015) Life cycle assessment of beef cattle production in two typical grassland systems of southern Brazil. J Clean Prod 96:426–434
32. Duyan Ö, Ciroth A (2013) Life cycle costing quick explanation: two different methods to perform life cycle costing in openLCA. GreenDelta, Berlin
33. Eady S, Viner J, MacDonnell J (2011) On-farm greenhouse gas emissions and water use: case studies in the Queensland beef industry. Anim Prod Sci 51(8):667–681

34. Efroymson RA, Dale VH, Langholtz MH (2017) Socioeconomic indicators for sustainable design and commercial development of algal biofuel systems. Renew Energy 9:1005–1023
35. EU (2014) How can we move towards a more resource efficient and sustainable food system. European Union, Brussels
36. EY (2018) Investor's guide to the queensland beef supply chain. Ernest & Young Australia Operations Brisbane
37. Flachowsky G, Meyer U, Südekum K-H (2018) Resource inputs and land, water and carbon footprints from the production of edible protein of animal origin. Arch Anim Breed 61(17–36)
38. Flores G, Hoffmann D, Rostron L, Shorten P (2014) Solar plus storage the key to solar-generated savings for a feedlot in the Central West. Australian Goverment Department of Industry, Canberra
39. Florindo TJe, Florindo GIBdM, Talamini E, Costa JSd, Ruviaro CF (2017) Carbon footprint and life cycle costing of beef cattle in the Brazilian midwest. J Clean Prod 147, 119–29
40. Forster S-J (2018) Feed consumption and liveweight gain, Future Beef2018, https://futurebeef.com.au/knowledge-centre/beef-cattle-feedlots-feed-consumption-and-liveweight-gain/
41. Gerber PJ, Mottet A, Opio CI, Falcucci A, Teillard F (2015) Environmental impacts of beef production: review of challenges and perspectives for durability. Meat Sci 109:2
42. Gerber PJ, Steinfeld H, Henderson B, Mottet A, Opio C, Dijkman J, Falcucci A, Tempio G (2013) Tackling climate change through livestock—a global assessment of emissions and mitigation opportunities. Food and Agriculture Organization of the United Nations, Rome
43. Gnansounou E, Raman JK (2016) Life cycle assessment of algae biodiesel and its co-products. Appl Energy 161:300–308
44. Goedeken MK (2017) Cultivation of Chlorella Sorokiniana Using Beef Feedlot Runoff Holding Pond Effluent, Master of Science thesis, University of Nebraska
45. Gontard N, Sonesson U, Birkved M, Majone M, Bolzonella D, Celli A, Angellier-Coussy H, Jang GW, Verniquet A, Broeze J, Schaer B, Batista AP, Sebok A (2018) A research challenge vision regarding management of agricultural waste in a circular bio-based economy. Crit Rev Environ Sci Technol 48(6):614–654
46. GonzáLez-GonzáLez LM, Correa DF, Ryan S, Jensen PD, Pratt S, Schenk PM (2018) Integrated biodiesel and biogas production from microalgae: towards a sustainable closed loop through nutrient recycling. Renew Sustain Energy Rev 82(P1):1137–1148
47. GRAAGC (2016) Reducing greenhouse gas emissions from livestock: Best practice and emerging options. Global Research Alliance on Agricultural Greenhouse gases, Auckland
49. GRDC (2015) Farm Gross Margin 2015: a gross margin template for crop and livestock enterprises. Grains Research & Development Corporation, Adelaide
50. Griffin G, Batten D, Campbell PK (2013) The costs of producing biodiesel from microalgae in the Asia-Pacific region. Int J Renew Energy Dev 2(3):105–113
51. GRSB (2014) Draft principles & criteria for global sustainable beef, global roundtable for sustainable beef. Overijssel
52. Hunkeler D, Lichtenvort K, Rebitzer G (eds) (2008) Environmental life cycle costing. Society of Environmental Toxicology and Chemistry, New York
53. Kannan N, Saleh A, Osei E (2016) Estimation of energy consumption and greenhouse gas emissions of transportation in beef cattle production. Energies (Basel) 9(11):960
54. Lovrenčec L (2010) Highlights of socio-economic impacts from biogas in 28 target regions European Union, Brussels
55. Lupo CD, Clay DE, Benning JL, Stone JJ (2013) Life-cycle assessment of the beef cattle production system for the northern great plains, USA. J Environ Qual 42(5):1386–1394
56. Madeira MS, Cardoso C, Lopes PA, Coelho D, Cláudia A, Bandarra NM, Pratesa OAM (2017) Microalgae as feed ingredients for livestock production and meat quality: a review. Livest Sci 205:111–21
57. Manik YBS, Leahy J, Halog A (2013) Social life cycle assessment of palmoil biodiesel: a case study in Jambi Province of Indonesia. Int J Life Cycle Assess 18:1386–1392
59. Martin P (2016) Cost of production: Australian beef cattle and sheep producers 2012–13 to 2014–15. Australian Bureau of Agricultural and Resource Economics and Sciences, Canberra

60. Martínez-Blanco J, Lehmann A, Muñoz P, Antón A, Traverso M, Rieradevall J, Finkbeiner M (2014) Application challenges for the social Life Cycle assessment of fertilizers within life cycle sustainability assessment. J Clean Prod 69:34–48
61. McCabe A, Halog A (2016) Exploring the potential of participatory systems thinking techniques in progressing SLCA. Int J Life Cycle Assess
62. McGinn SM, Chen D, Loh Z, Hill J, Beauchemin KA, Denmead OT (2008) Methane emissions from feedlot cattle in Australia and Canada. Aust J Exp Agric 48
63. McKiernan B, Gaden B, Sundstrom B (2007) Dressing percentages for cattle. New South Wales Department of Primary Industries, Sydney
64. MLA (2012) National guidelines for beef cattle feedlots in Australia. Meat & Livestock Australia, Sydney
65. MLA (2015) Beef cattle nutrition: an introduction to the essentials. Meat & Livestock Australia, Sydney
66. MLA (2017) Cattle assessment manual. Meat & Livestock Australia, Sydney
67. Mobin S, Alam F (2014) Biofuel production from Algae utilizing wastewater, paper presented to 19th Australasian Fluid Mechanics Conference, Melbourne, 8–11 December
68. Moss J, Coram A, Blashki G (2014) Solar energy in Australia: health and environmental costs and benefits, The Australian Institute, Canberra
69. Murinda SE (2013) Algae for conversion of manure nutrients to animal feed: evaluation of advanced nutritional value, toxicity, and zoonotic pathogens. USDA, Washington
70. Noi CD, Ciroth A, Srocka M (2017) OpenLCA 1.7: Comprehensive user manual. GreenDelta, Berlin
71. Norris CB, Norris GA, Aulisio D (2014) Efficient assessment of social hotspots in the supply chains of 100 product categories using the social hotspots database. Sustainability 6:6973–6984
72. Ogino A, Kaku K, Osada T, Shimada K (2004) Environmental impacts of the Japanese beef-fattening system with different feeding lengths as evaluated by a life-cycle assessment method1. J Anim Sci 82(7):2115–2122
73. OpenLCA (2017) OpenLCA 1.7.0.beta, GreenDaelta, Berlin
74. Pagotto M, Halog A, Costa DFA, Lu T (2021) A sustainability assessment framework for the Australian food industry: integrating lifer cycle sustainability assessment and circular economy. In: Muthu SS (ed) Life cycle sustainability assessment, Springer Nature
75. Pagotto M, Halog A (2016) Towards a circular economy in australian agri-food industry: an application of input-output oriented approaches for analyzing resource efficiency and competitiveness potential. J Ind Ecol 20(5):1176–1186
76. Peters GM, Rowley HV, Wiedemann S, Tucker R, Short MD, Schulz M (2010) Red Meat production in Australia: life cycle assessment and comparison with overseas studies. Environ Sci Technol 44(4):1327–1332
77. PHIDU (2018) Monitoring inequality in Australia: Queensland, Adelaide
78. Provolo G, Mattachini G, Finzi A, Cattaneo M, Guido V, Riva E (2018) Global warming and acidification potential assessment of a collective manure management system for bioenergy production and nitrogen removal in Northern Italy. Sustainability (Basel, Switzerland) 10(10):3653
79. Ridoutt BG, Sanguansri P, Freer M, Harper GS (2012) Water footprint of livestock: comparison of six geographically defined beef production systems. Int J Life Cycle Assess 17(2):165–175
80. Robson S (2007) Beef cattle vaccines. New South Wales Department of Primary Industries, Waga Waga
81. Schenk P (2016) On-farm algal ponds to provide protein for northern cattle. Meat and Livestock Australia Limited, Sydney
82. Schleiss K, Jungbluth N (2018) Life cycle inventories of bioenergy. Ecoinvent, Zurich
83. Skiba U, Rees B (2014) Nitrous oxide, climate change and agriculture. CAB Rev 9:1–7
84. SWA (2018) Statiscal Tables, Safe Work Australia, Canberra, 20 Jan 2018, https://www.safeworkaustralia.gov.au/resources_publications/Statistical-tables

85. Swarr TE, Hunkeler D, Klöpffer W, Pesonen H-L, Brent AC, Pagan R (2011) Environmental life-cycle costing: a code of practice. Int J Life Cycle Assess 16:389–91
86. Symeonidis A, Levova T, Ruiz EM (2018) *Anaerobic digestion of manure—GLO*, Ecoinvent Centre, Zurich
87. Thomson J (2019) Beef cattle farming in Australia. IBISWorld, Sydney
88. Traverso M, Asdrubali F, Francia A, Finkbeiner M (2012) Towards life cycle sustainability assessment: an implementation to photovoltaic modules. Int J Life Cycle Assess 17(8):1068–1079
89. Treyer K, Vadenbo C (2018) Electricity production, photovoltaic, 570kWp open ground installation, multi-Si—AU, Ecoinvent, Zurich
90. Tucker R, McDonald S, O'Keefe M, Craddock T, Galloway J (2015) Beef cattle feedlots: waste management and utilisation. Meat & Livestock Australia, Sydney
91. UNEP (2009) Guidelines for social life cycle assessment of products. United Nations Environment, Kenya
92. UNEP (2011) Towards life cycle sustainability assessment. United Nations Environment Programme, Brussels
93. UNEP (2013) The methodological sheets for subcategories in social life cycle assessment (S-LCA). United Nations Environment Programme, Brussels
94. UNEP (2015) Towards a life cycle sustainability assessment. United Nations Environment, Kenya
95. Vries Md, Boer IJMd (2010) Comparing environmental impacts for livestock products: a review of life cycle assessments. Livest Sci 128:1–11
96. Walsh B, Rydzak F, Palazzo A, Kraxner F, Herrero M, Schenk P, Ciais P, Janssens I, Peñuelas J, Niederl-Schmidinger A, Obersteiner M (2015) New feed sources key to ambitious climate targets. Carbon Balanc Manag 10(1):1–8
97. Watkins M, Castlehouse H, Hannah M, Nash DM (2011) Nitrogen and phosphorus changes in soil and soil water after cultivation. Appl Environ Soil Sci 2:1–10
98. Watts PJ, Davis RJ, Keane OB, Luttrell MM, Tucker RW, Stafford R, Janke S (2016) Beef cattle feedlots: design and construction. Meat & Livestock Australia, Sydney
99. Watts P, McCabe B (2015) Feasibility of using feedlot manure for biogas production. Meat & Livestock Australia, Sydney
100. WGEA (2018) Gender equity insights 2017: inside Australia's Gender Pay Gap Workplace Gender Equality Agency, Sydney
101. Wiedemann SG, Henry BK, McGahan EJ, Grant T, Murphy CM, Niethe G (2015) Resource use and greenhouse gas intensity of Australian beef production: 1981-2010. Agric Syst 133:109–18
102. Wiedemann SG, Murphy CM, McGahan EJ, Bonner SL, Davis R (2014) Life cycle assessment of four southern beef supply chains. Meat & Livestock Australia, Sydney
103. Wiedemann S, McGahan E, Murphy C, Yan M-J, Henry B, Thoma G, Ledgard S (2015) Environmental impacts and resource use of Australian beef and lamb exported to the USA determined using life cycle assessment. J Clean Prod 94:67–75
105. Wiedemann S, Davis R, McGahan E, Murphy C, Redding M (2017) Resource use and greenhouse gas emissions from grain-finishing beef cattle in seven Australian feedlots: a life cycle assessment. Anim Prod Sci 57(6):1149–1162
106. Wiedemann SG, McGahan EJ, Watts PJ (2010) Scoping life cycle assessment of the Australian lot feeding sector. Meat & Livestock Australia, Sydney
107. Yang Y, Zhang B, Cheng J, Pu S (2015) Socio-economic impacts of algae-derived biodiesel industrial development in China: an input–output analysis. Algal Res 9:74–81
108. Yasar A, Nazir S, Tabinda AB, Nazar M, Rasheed R, Afzaal M (2017) Socio-economic, health and agriculture benefits of rural household biogas plants in energy scarce developing countries: A case study from Pakistan. Renew Energy 108:19–25
109. Yazan DM, Cafagna D, Fraccascia L, Mes M, Pontrandolfo P, Zijm H (2018) Economic sustainability of biogas production from animal manure: a regional circular economy model. Manag Res Rev 41(5):605–624

110. Yu M, Halog A (2015) Solar photovoltaic development in Australia-a life cycle sustainability assessment study. Sustainability (Switzerland) 7(2):1213–1247
111. Zijp MC, Heijungs R, Voet E, Meent D, Huijbregts MAJ, Hollander A, Posthuma L (2015) An identification key for selecting methods for sustainability assessments. Sustainability 7:2490–2512

Life Cycle Sustainability Assessment Study of Conventional and Prefabricated Construction Methods: MADM Analysis

Ali Tighnavard Balasbaneh, David Yeoh, and Mohd Irwan Juki

Abstract Lately, many governments have been significantly promoting modular building instead of conventional as a practical solution toward enhancing sustainability in the construction sector. Therefore, this research aims to compare traditional and modular building construction to find each environmental and cost difference as a criterion for comparison. This study's life cycle sustainability assessment comprises embodied energy, greenhouse gas (GHG), and cost. The result showed that the steel modular has the lowest embodied energy and carbon emission following conventional steel construction. For traditional construction, 28% of GHG emissions are related to on-site activity, while PPVC is less than 1%. However, the development of the factory is about 11% of the total construction emission for PPVC. On the other hand, the concrete, conventional method has a lower construction cost following by concrete modular. The transportation cost of modular building is responsible for up to 13% of the total construction cost. While the conventional building has a higher worker wage by 11%, compare to modular construction. Multi-attributes decision-making (MADM) using WASPAS has been applied to reveal the best construction material and method. The result showed that steel modular is the best option for construction.

Keywords Modular construction · Conventional construction · Environmental assessment · Embodied energy · Greenhouse gas

A. T. Balasbaneh (✉) · D. Yeoh · M. I. Juki
Faculty of Civil and Environmental Engineering, Universiti Tun Hussein Onn Malaysia, 86400 Parit Raja, Johor, Malaysia
e-mail: tighnavard@uthm.edu.my

D. Yeoh
e-mail: david@uthm.edu.my

M. I. Juki
e-mail: juki@uthm.edu.my

S. S. Muthu (ed.), *Life Cycle Sustainability Assessment (LCSA)*,
Environmental Footprints and Eco-design of Products and Processes,
https://doi.org/10.1007/978-981-16-4562-4_8

1 Introduction

There's an unusual request available around the world for constructing the new buildings, either residential or commercial [36], which this demand also very high in Malaysia [19]. The construction sector's importance is not limited to its environmental impact and also must give attention to its economic aspect because it contributed to 12% of the worldwide economy and contribution more than 110 million construction workers in this industry [2]. Meanwhile, the construction sector uses 50% of the global resources worldwide [1]. The building industry produces more than forty percent of carbon emissions and thirty-five percent of wastage to the environment yearly [38]. Thereby, evaluation of the building system still seems vital to decreasing any opportunity from this sector. Meantime, these statistics can increase in a few decades by increasing the population. The construction buildings sector is responsible for high energy usage and carbon emission in their construction and manufacturing stage [47].

Lately, by promoting new sustainable construction technique called Prefabricated Prefinished Volumetric Construction (PPVC) has been introduced and encouraged by some governments around the world. Modular or knows as PPVC in some countries, it has been increasing every year in Asian countries such as Malaysia, Hong-Kong, and Singapore [8]. Off-site construction (modular or PPVC) is a technique that builds elements and components off-site and all processes has accomplished in the factory and then transports it to the site for installation [45]. On the other hand, the conventional construction process has been manufacturing on-site. Modular, unlike traditional construction methods where works are executed consecutively. Although there is much research accomplished on off-site construction [16, 29], only a few had explored PPVC. Therefore, this research has compared concrete and steel (Light steel framework) by two different construction methods, namely, on-site and modular, to find out its pros and cons. PPVC has some advantages, such as producing less wastage and using fewer workers on-site, but it comes with some disadvantages [28, 34]. Some of the obstacles are related to the increased logistics and transportation and higher initial cost which make it a sensitive issue for housemakers to use PPVC techniques in their projects. Regardless of the materials that are chosen for their construction. Therefore, assessing PPVC and comparing it with conventional construction (still very famous in many countries) is vital.

Environment evaluation is one of the most critical dimensions of sustainable construction assessment [9], especially for PPVC. Tavares et al. [52] accomplished a study about GHG of modular building within the construction phase (cradle-to-gate). They believed the transportation might be challenging for sending modules for abroad construction. Quale et al. [48] compared wooden PPVC with conventional buildings within production and construction stages. The result showed that modular construction had a lower impact compared to the traditional structure. Monahan and Powell [43] revealed that off-site construction techniques and selecting the proper building materials could save embodied energy. Previous research believed that modular construction manufacturing could decrease GHG emissions by 3% over a 50-year [32]. Thormark [55] suggested changing building materials can lower the embodied energy up to 17%.

Guggemos and Horvath [22] compared concrete and steel office buildings. The result showed that the steel frame has a high embodied energy compared to concrete in the total life cycle stage by excluding the maintenance phase. Zhang et al. [61] indicated that concrete framed has lower environmental emissions and energy consumption than a steel frame. Alshamrani [5] reported that concrete structures have a lower environmental impact compared to steel-framed buildings.

Gustavsson and Joelsson [23] believed that energy consumption is different for each construction technique. For example, the primary energy production for low-energy and conventional residential buildings is 60 and 45% of the total energy consumption, respectively. Heravi et al. [26] assessed the energy consumption of concrete and steel frame buildings limited to production and construction phases. The result showed that the energy consumption of concrete structures is approximately 27% less than the steel structure. Han et al. [24] analyzed the energy consumption of buildings, and results revealed that steel and cement have the highest contribution to energy consumption. Xing et al. [57] compared environmental emissions and energy consumption of steel and concrete frames. The result indicated that the energy consumption of steel manufacturing is higher than concrete.

Some studies had a contradictory result in comparison between concrete and steel buildings. Mao et al. [41] assessed the environmental emissions of off-site prefabrication and conventional constructions methods. The assessment showed that concrete prefabricated structures produce higher emissions compared to prefabricated steel constructions. Similarly, [57] compared the life cycle assessment (LCA) for steel and concrete for office buildings. The results suggest that concrete exhibiting a higher cost than steel. However, opinions on the extent to which LCA reduction can be achieved through prefabrication are still inconsistent. Caruso et al. [15] carried out an LCA-based comparison of the environmental impact of building materials. The result showed that concrete structure has a more significant influence on carbon emission compared to steel. Some research emphasizes the importance of conducting CO_2 emission for buildings [49]. Li et al. [35] assessed the LCA of CO_2 for precast concrete during the construction stage. Results showed that carbon emissions of construction machinery reached 73% of the total during the construction phase. Teng [53] assessed the carbon emissions of prefabricated buildings. The result revealed that 15.6% of embodied carbon was achieved through prefabrication compared to traditional base cases.

Although some studies (e.g., [46] had shown that steel buildings have a higher carbon emission than concrete buildings, some other research (e.g., [61] showed that concrete structures produce higher emissions. These variables make it difficult to compare the reported results of carbon emissions in a meaningful way, thus leading to a vast disparity in the works [45].

The economic factor is one of the significant concerns of construction companies worldwide [11]. Kamali and Hewage [30] believed that researchers mainly concern about the environmental performance of modular. Therefore, neglect the economic aspect of PPVC compared to conventional buildings. Ho et al. [51] believed that modular has a higher cost than traditional construction by up to 20% and 25%, respectively, related to the concrete and steel PPVC. Much research in literature emphasizes that using off-site construction is still relatively small [7, 21, 60]. This research compares the two most relevant materials for construction, such as concrete and steel, with two different construction methods, namely modular and conventional, due to their environmental and cost aspects.

As a subset of sustainable development, sustainable construction is significant because half of the total raw material extracted from the planet is used in construction. More than half of the waste comes from the construction sector [44]. However, how much this can affect the environmental and cost aspects in the construction sector remains unanswered for modular compared to conventional construction for both concrete and steel. In general, there is a lack of evidence of references on concrete and steel PPVC in the Malaysian construction industry due to construction challenges and external factors in the adoption of new construction technology. Therefore, builders still utilize conventional building strategies despite the uncertainty of which method is more dominant. Thus, the life cycle sustainability assessment has been applied by considering the embodied energy, climate change, and cost evaluations for different construction materials and techniques in this research.

2 Methodology

Table 1 shows the system boundaries correspond to the life cycle modules A1, A2, A3, A4, and A5 of the EN15804 and EN15978 standards [18]. Hence, according to EN15978, the system boundary of the analysis entails the stage of material production and construction stage, the use stage (maintenance), the end-of-life stage. Finally, the last step allocated the benefits and loads due to recycling and reuse materials. The research comprises the cradle-to-gate boundary. The first stage is related to the material phase, and it is divided into energy requirements for raw material extraction

Table 1 A modular building's life cycle stages from EN15804 and EN15978

Product stage			Construction stage		Use stage							End of life				Benefit and loads			
A1	A2	A3	A4	A5	B1	B2	B3	B4	B5	B6	B7	C1	C2	C3	C4	D1	D2	D3	D4
Raw Material Extraction	Transport	Manufacturing in Factory	Transport To Site	Installation and Erected Process	Use	Maintenance	Repair	Replacement	Refurbishment	Operation Energy Use	Operation Water Use	Demolition	Transport	Waste Processing	Disposal	Reuse	Recovery	Recycling	Landfill
X	X	X	X	X															

(A1), transportation to factory (A2), and manufacturing process (A3). The phase is the construction stage that contributed to transporting materials or components to the site (A4) and processing of installation (A5).

2.1 *Life Cycle Assessment*

The LCA methodology evaluates the analysis of inputs and outputs for each modular structure at each stage of the product's life cycle. LCA is the principal methodology for assessing environmental impacts related to the product evaluation during its all life cycle stages. As already mentioned above, this stage aims to utilize LCA [33] for two sophisticated parameters, namely, embodied energy and GHG emission, to characterizing the environmental impact of a distinctive modular.

Embodied energy and environmental impacts associated with the construction materials have been investigated with the help of SimaPro software. LCA needs efficient data collection for each stage of product such as manufacturing, maintenance, and end-of-life stage for the emission were generated during each step of the product's life [6]. The LCA study needs to follow ISO 14,040 [25] to validate research and reliable results. LCA approach has investigated the "cradle-to-grave" of process or product during the life cycle phase. The other important issue in any LCA is the functional unit that needs to define correctly. The available unit has been described in this research as one m^2 of gross floor area.

2.2 *Life Cycle Inventory and Life Cycle Impact Assessment*

Two main components of any LCA are related to the life cycle inventory (LCI) and life cycle impact assessment (LCIA). LCI analysis involves collecting the required data and calculating the related inputs and outputs using the LCI database. The Ecoinvent database is a national factor that allocates procedures to all materials [20]. However, the Ecoinvent used in the current research was adjusted according to the Malaysian situation by applying the local power mix information as recommended by Horv et al. [27] and investigated by Balasbaneh et al. [9] to achieve a more accurate result.

LCIA is performed based on the results of the inventory investigation to determine the environmental impacts of buildings [25]. The LCIA organizes a set of potential environmental impact assessment measures embodied energy and GHG. Carbon impact [14] value, which is evaluated beside Embodied energy, relates to the accumulated GHG caused by an item amid its life cycle. LCA evaluates and reports GWP to indicate the degree to which a building, over its lifetime, may contribute to climate change. Although assessing energy consumption is vital in finding the area that reduction related to consumption of material or product could be achieved, it has not necessarily delivered the proper indication of consumption associated with environmental impact. The GHG emission has been recognized as a helpful indicator cover over all effects from the environment, and kg CO_2eq represents it. GHG is calculated based on IPCC (Intergovernmental Panel on Climate Change) method for 100 years (IPCC 2007). Carbon dioxide emissions are defined as the emissions that are released directly and indirectly from manufacturing materials or products over their lifetime [37]. Different fuel resources for material products such as renewable resources and coal result in various impacts on GHG or embodied energy. In Malaysia, the electricity consumed in producing materials such as concrete prefabricated is considered fossil fuel consumption apportioned as 96.63% of the national electricity in 2018. The electricity generation in Malaysia [13] is different from other countries since fossil fuel is the primary source of production [54]. A carbon footprint is defined as "the total amount of carbon dioxide emissions, directly and indirectly, caused by an activity or that accumulated over a product lifetime" [37].

The second environment indicator is embodied energy that contributed to the energy-consuming consumption for the total life cycle of products from raw material acquisition to disposal stage. It is necessary to incorporate it into the material itself. It has assessed the nonrenewable primary energy and excluded the renewable energy consumed for production and construction [40]. Embodied energy is related to the energy consumed in the manufacturing of concrete, steel, etc., used during this process. For example, concrete comprises energy for material extraction to produce cement, transporting and processing in the factory to produce cement and deliver to the factory to construct the modular.

There are various factors such as fuel supply, different regions of study, technology, and analysis method that can affect the result. The assessment for the modular building has applied an input–output-based hybrid analysis in this study [56]. Embodied energy is measured as the quantity of nonrenewable energy per unit of building material, component, or system. Embodied energy values have been calculated in SimaPro using the Cumulative Energy Demand (CED) method 1.04. Some of the materials are used in more than one form. It is expressed in megajoules (MJ) per unit weight (kg or ton).

2.3 *Life Cycle Costing*

Sustainability in a holistic view is required for balance between environment and cost for the construction sector. However, one of the main criteria for a successful construction project is to manage the cost. In the current case studies, life cycle costing (LCC) was performed by an Excel spreadsheet for 50 years' costs of a flooring system. Estimation is based on the standard construction cost guide handbook [17] and National Construction Cost Centre (CIDB Malaysia Official Portal) in Malaysian Ringgit (RM). Applying the cost perspective in the early design stages of construction leads to a better understanding of the cost to decision-makers to ensure the best choice among different materials. The building sector is a long-term investment with potential environmental impacts [11]. Recent research into the cost of building in

Malaysia [3] shows that most construction projects start without implementing a proper LCC assessment in their management, mainly related to their commitment and policy. Five major LCC elements were assessed, namely: material, wages, transportation, maintenance, and end of life for each alternative floor system based on the construction cost data from the official portal of the Department of Statistics Malaysia and Malaysia official portal (CIDB). The current research assesses cost based on the 2018 Malaysian ringgit present value (PV). LCC following the equation below (1):

$$\text{LCC} = \text{Initial Cost} + \text{PV\_Maintenance Cost} + \text{PV\_Demolition Cost} \quad (1)$$

The assumption of this study is to assess the cost based on 3.4% Inflation Rate (Malaysia Inflation Rate data), which follows the Malaysia ringgit (PV) 2018, 4.5% discount rate (Malaysia formal Discount Rate data), 38.53 Sen/kWh electricity cost (Malaysiakini data), and finally 0.31 MYR for each ton per kilometer for the cost of lorry transportation following previous research about the cost of building material by [13]. One Malaysian ringgit is equal to 0.25 United States dollars (December 2018).

2.4 *Multi-attributes Decision-Making*

The first step entails the weighting calculation for each criterion. The survey was accomplished using the AHP method and Saaty's evaluation. The experts included a design engineer, a professor/university lecturer, and a construction manager. The AHP method depends on the discernment of the specialists and the pairwise comparison of the criteria; the preferred solution must be agreed upon with the interest groups who usually have different goals. The relative significance scale ranged between one and nine. The inclination scale for the pairwise comparison of two parameters ranged from the most extreme of 9–1. The scoring system for the pairwise comparison of elements in the hierarchy is: 1 = Indifferent, 3 = Low priority, 5 = Moderate priority, 7 = Strong priority, and 9 = Very Strong priority. To ensure consistency of the comparison framework, the computed Consistency Ratio (CR) should be less than 0.1. The following describes the step by step calculation of this method. The criteria and their corresponding weights (w) that need to be placed in the pairwise comparison matrix are in line with Saaty [50], i.e., with GHG (X1), embodied energy (X2), and Economics (X3).

Weighted Aggregated Sum Product Assessment (WASPAS) method was developed by [59]. It is a group of two methods: Weighted Sum Method (WSM) [39] and Weighted Product Method (WPM). The detailed procedure of WASPAS method is as follows:

Step 1: Initialize the matrix for solving the selection problem.

Step 2: Normalize the decision matrix as shown in Eqs. (2) and (3).

$$x_{ij} = \frac{x_{ij}}{\max x_{ij}} \tag{2}$$

$$x_{ij} = \frac{x_{ij}}{\min x_{ij}} \tag{3}$$

where x is the assessment values. Equations (1) and (2) are used for maximization (beneficial) and minimization (no beneficial) criteria, respectively.

Step 3: Calculate the total relative importance based on the WSM method using Eq. (4).

$$Q_i^{(1)} = \sum_{j=1}^{n} x_{ij}.w_j \tag{4}$$

Step 4: Calculate the total relative importance based on the WPM method using Eq. (5).

$$Q_i^{(2)} = \Pi_{j=1}^{n} (x_{ij})^{w_j} \tag{5}$$

Step 5: The rank order accuracy is helpful in the decision-making process, in the WASPAS method, a more general equation for formative and the total relative significance of alternatives is given by Eq. (6).

$$\boldsymbol{Q}_i = \lambda.\boldsymbol{Q}_i^{(1)} + (1 - \lambda).\boldsymbol{Q}_i^{(2)} \tag{6}$$

where λ value is considered as 0.5.

3 Case Study

The selected case studies include the single-story residential building. Two different construction techniques will allow comparative analysis to concentrate on sustainability assessment, namely, economic and environmental. The dimension of the case study is $12 \times 7.5\ m^2$. The general data information about case studies is available in Table 2. The first stage for on-site construction is related to the material production and Construction phase (A1–A5): the phase consisting of material extraction, material transportation, construction and installation, and product and worker transportation. The on-site machinery such as concrete pump, concrete mixer, gas engine vibrator, and disc cutter has been considered for OSC on site. The density of concrete for all three case studies is 2400 kg/m^3. Figure 1 shows the architectural plan of the case study that has been assessed in this research.

Modular construction (PPVC) includes pre-assembled room-sized volumetric units that are ordinary, wholly fitted out in manufacture, and are installed on location as load-bearing 'building blocks.' It has a hollow core on each side of the wall. The

Table 2 Characterization of construction materials

Buildings	Material	Unit	Thickness	Weight per kg/M	Transport to Factory	Total weiglit/kg
Concrete	Precast concrete	mm	200	291	Lorry 32 ton, 30 krn	52380
	Reinforcing steel	Diameter/mm	18	26.4	Lorry 3.5 ton, 7 bn	1964.8
	Tile floor	mm	5	6.4	Lorry 16 ton, 25km	675
	Mineral wool insulation	MM	21	2.3	Lorry 16 ton, 25km	675
Steel	Steel stud	MM	200	4.9	Lorry 16 ton, 25km	4650
	Tile floor	MM	20	6.4	Lorry 16 ton, 25km	675
	Polyethylene pipe	Diameter/mm	110	3.2	Lorry 3.5 ton, 7 km	64
	Mineral wool insulation	MM	21	2.3	Lorry 16 ton, 25km	190

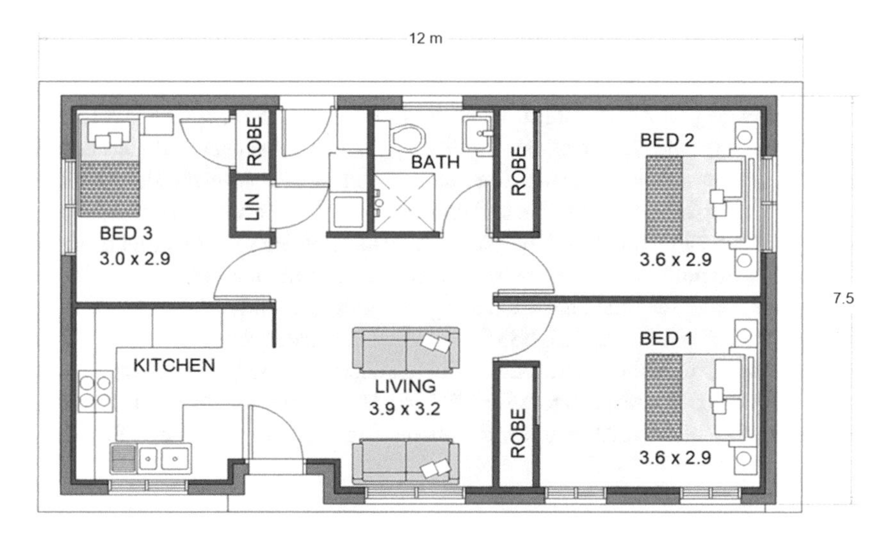

Fig. 1 Architectural plan of case study

starter bar comes from underneath the slab and passes through the hollow center. The first stage of PPVC is related to the raw material extraction, transport, and manufacturing of wall, roof, and slab in the Factory (A1–A3). In the next stage, the panel sends to the site, and by using of crane, the assembly will process (A4–A5), and the panel transported about 150 km from the construction site.

4 Results

4.1 Embodied Energy

The production phase of concrete is responsible for more embodied energy than steel, as shown in Table 3. However, both steel modular and conventional production is the same due to the same building plan. That is valid for concrete modular and traditional as well. The embodied energy for construction in the factory is only applicable for modular PPVC, and concrete has a slightly higher energy usage in this sector. Comparison of waste concrete PPVC has shown more impact compare to steel PPVC. In the meantime, the waste for PPVC modular is about 2% of total embodied energy, contributing to 5% for conventional construction. That shows that PPVC construction can reduce the amount of construction waste. Transportation is one of the critical aspects of PPVC construction. The result indicates that the transport of units to the site for PPVC is higher than conventional construction. In practice, the embodied energy of PPVC transportation is 45% higher than when the use of traditional construction. That is due to the higher distance for PPVC of factory distance from the site. Both PPVC and conventional construction contributed to the embodied energy of on-site construction impact. Despite the fact that PPVC has a lower impact on the use of cranes for lifting the unit. However, still, it needs to consider getting a reliable result. The construction on site is responsible for 12% and 9%, respectively, for concrete and conventional steel building. However, if the factory is responsible for 10 and 9% for steel and concrete PPVC, construction shows a lower contribution to embodied energy than the conventional method.

Figure 2 shows the total embodied energy of each case study. The total embodied energy of both steel techniques is 383 GJ and 395 GJ, respectively, for steel PPVC and steel conventional method. The total embodied energy for both concrete techniques is 491 GJ and 505 GJ, respectively, for PPVC and traditional construction. The result revealed that steel has relatively lower embodied energy than concrete in PPVC and

Table 3 Embodied energy

Embodied energy	Unit	Production	Construction in factory	Waste	Transport	Construction on-site
S-PPVC	GJ	330.24	40	6	5.58	1.58
S-C		330.24	–	16	2.5	46.4
C-PPVC		431.3	45	8	5.58	1.58
C-C		431.3	–	25	2.5	47

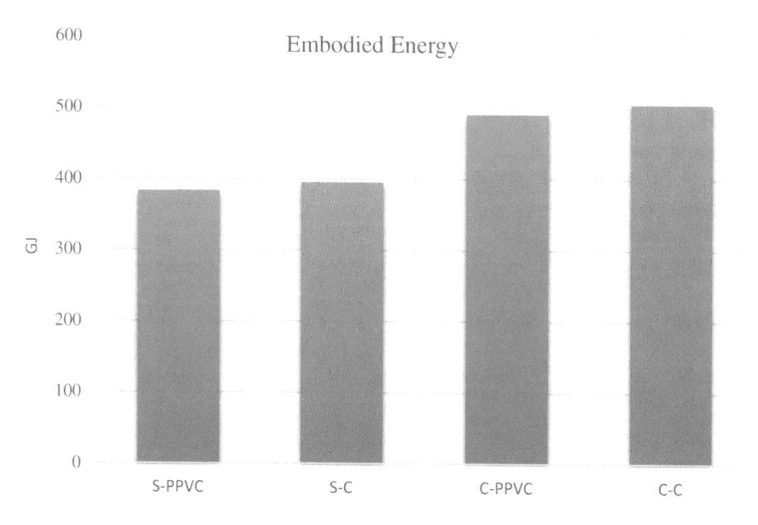

Fig. 2 Embodied energy, modular steel (S-PPVC), conventional steel (S-C), modular concrete (C-PPVC), and conventional concrete (C-C)

conventional construction methods. Construction steel building by modular building contributed to 3% lower embodied energy than traditional construction of steel. On the other hand, concrete PPVC also has a 3% lower embodied energy than a conventional concrete building. However, in comparing embodied energy between concrete and steel, the result shows the concrete has 22% higher embodied energy.

4.2 *Greenhouse Gas Emission*

The result of the LCA on GHG for different life cycle phases is shown from cradle-to-gate in Table 4. The result indicates that the production stage contributed to higher emissions for both PPVC and conventional construction methods. On the other hand, PPVC for concrete and steel has lower GHG emissions in waste and construction on-site activity. However, the total emission for conventional construction is higher than PPVC due to transporting equipment and machinery to the construction site and

Table 4 Greenhouse gas emission

GHG	Unit	Production	Construction in factory	Waste	Transport	Construction on-site
S-PPVC	Kg CO_2eq	16,480	2284	226	980	190
S-C		16,480	–	780	320	6720
C-PPVC		26,360	4360	510	980	190
C-C		26,360	–	1220	320	7200

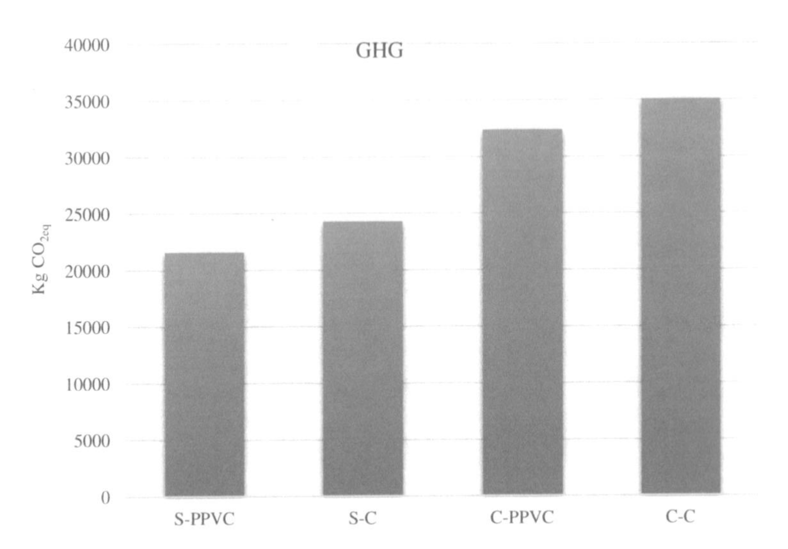

Fig. 3 Greenhouse gas emission, modular steel (S-PPVC), conventional steel (S-C), modular concrete (C-PPVC), and conventional concrete (C-C)

wastage. The total emission of construction for steel PPVC is about 12% less than conventional steel construction (20,160 kg CO_2eq vs. 24,300 kg CO_2eq). While this percentage is different for concrete and C-PPVC emission is about 8% compared to traditional concrete construction.

However, waste for constructing the modular building is lower than conventional construction for both steel and concrete materials. It should state that the waste for PPVC is about only 1%, while it exceeds 3% for traditional construction. Another essential category between PPVC and conventional techniques is related to their transportation diversity. The emission related to transportation from PPVC is about 33% higher than the traditional technique, and this is a drawback for modular PPVC. Figure 3 shows the final comparison between concrete and steel that have been used in conventional and modular construction. The result indicated that modular steel construction contributed to lower GHG emissions compared to alternatives from cradle-to-gate. The total CO_2 emission for steel is 21,600 kg CO_2eq and 24,300 kg CO_2eq, respectively, for S-PPVC and S-C. On the other hand, the total CO_2 emission for concrete is 32,400 kg CO_2eq and 35,100 kg CO_2eq, respectively, for C-PPVC and C-C. The result revealed that regardless of the steel construction method, it has a lower GHG emission than concrete.

4.3 Life Cycle Costing

Table 5 shows the material cost for both steel and concrete structures regardless of construction techniques. The higher cost of the steel structure is related to the light

Table 5 Cost estimation

Types	Material	Cost of construction per m^2 MYR	Total Cost M^2/MYR
Steel	Light Steel Stud Frame Walls and ceiling	94	169
	Plasterboards	35	
	Mineral wool	22	
	OSB sheathing boards	18	
Concrete	Concrete panels & slab	85	150
	Reinforcing bars at both faces of all walls	32	
	Plasterboards	35	

steel stud frame with 55% of each m^2 following by plasterboards and mineral wool. In concrete structures, the reinforcing bars have been applied to walls and ceiling, and the cost is estimated by 21% for each per m^2 after concrete.

The cost of the S-PPVC is 54,895 MYR, 8200 MYR, 1400 MYR, and 3545 MYR, respectively, related to the production and construction, transport to site, crane, and wages. The cost of S-C is 54,895 MYR, 1105 MYR, 4200 MYR, respectively, related to the production and construction, transport to site, and wages. The result shows that the transportation cost of PPVC is 86% higher than the conventional method. While, the cost of C-PPVC are 48,655 MYR, 8200 MYR, 1400 MYR, and 3545 MYR, respectively, related to the production and construction, transport to site, crane, and wages. On the other hand, the cost of C-C is 50,900 MYR, 1600 MYR, 6500 MYR, respectively, related to the production and construction, transport to site, and wages.

Figure 4 shows the total cost for each case studies and their related distribution based on percentages. The total cost of steel PPVC is higher than other alternatives by 68,040 MYR. The second highest is associated with the conventional steel method equal to 60,200 MYR. The result showed that steel structure either as modular or traditional has a higher total construction cost than concrete structure. The total cost of concrete is 61,800 MYR and 59,000 MYR, respectively, for PPVC and conventional methods. The modular construction cost is higher than traditional methods in both steel and concrete structures. Moreover, the transportation cost is responsible for 12 and 13% of total construction for modular, respectively, steel and concrete. In comparison, transportation is only accountable for 2% and 3%, respectively, for steel and concrete structures in conventional construction techniques.

The worker wages also show the different percentages in comparison between modular and conventional construction. Its cost is about 5% and 6%, respectively, for steel and concrete in modular building. While it even increases to 11% for a traditional concrete structure. Steel PPVC has shown the lowest contribution to cost-related production and construction (P and C). While, in the steel conventional method, it contributed to 91% of total construction cost (Fig. 5).

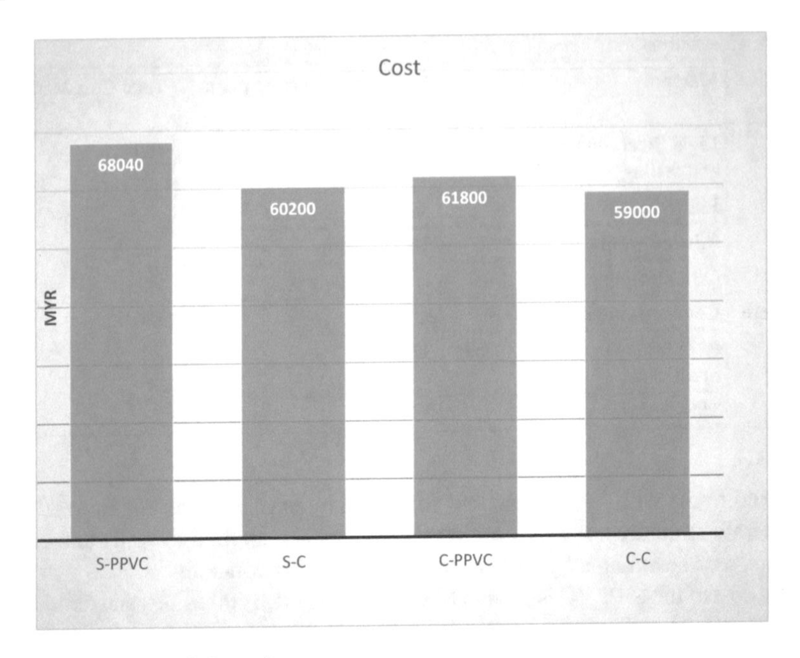

Fig. 4 Life cycle cost of alternatives

4.4 Multi-attributes Decision-Making

At this stage, MADM is chosen in the current research to find the best of the alternatives and provide a balance between the CO_2 emissions and embodied energy and an economic impact. Firstly, the importance of each criterion is evaluated using expert knowledge in the construction field. Generally, the decision-maker has to choose the best alternative among numerous alternatives by considering conflicting criteria. The importance ranges from 1 to 9, and the results are shown in Table 6 for all the criteria. A computed CRa of less than 0.1 indicates that the comparison matrix is consistent. Three groups assist in prioritizing the three different parameters of this research, and based on the Construction manager's opinion; the cost should be a priority. Based on the construction manager's view, GHG is allocated in the second parameter, while Embodied Energy is in the third priority.

On the other hand, Designers has a slightly different opinion on this matter and believe that Embodied Energy and GHG should be considered as a priority in choosing the suitable options. On the other hand, this expert believes that the cost is a different priority. The third group in this survey entails the Academic/Professor who believes that Embodied Energy should be considered the highest priority. As shown in Table 6, the Academic put the cost as second priority and GHG emission as lower priority. Finally, based on the cumulative results of the expert opinions, Embodied Energy is determined as the priority with 0.36 weighting followed by GHG with 0.35 weighting. Meanwhile, the cost settled as the last priority with weightings of 0.29.

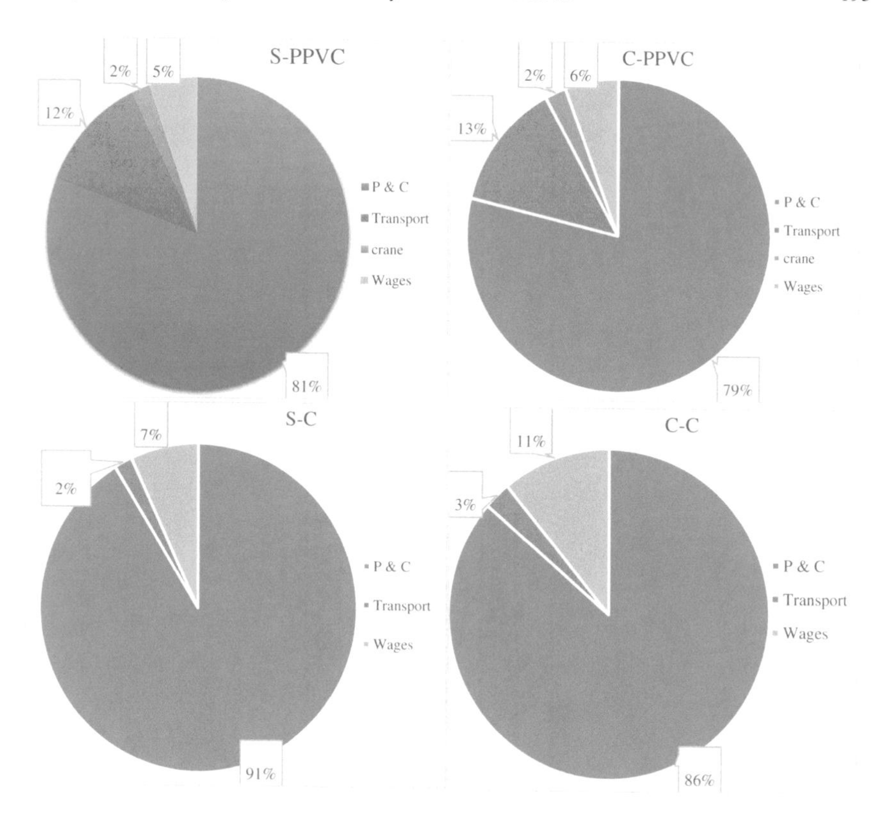

Fig. 5 Cost estimation, modular steel (S-PPVC), conventional steel (S-C), modular concrete (C-PPVC), and conventional concrete (C-C)

The initial decision-making matrix for the different construction methods and materials is presented in Table 7. This result originated from Tables 3, 4, and Fig. 3.

The dimension beneficial for embodied energy is the one that has a lower amount than equal to 383.4 (Table 7). That is also indefeasible for GHG and cost following by 21,600 and 59,000. Therefore, in all three criteria, the minimum amount is considered beneficial. In step 2, it needs to normalize the decision matrix using Eq. (2), and the assessment could be found in Appendix 1. The next step is calculating the relative importance method by implying the weighting of each criterion extracted from Table 6. The result of equation three has shown in Table 8 by the sum of all related criteria for each case study. In that case, the sum of 0.36, 0.35, and 0.25 (Appendix 2) will be 0.9615. Step 4 is related to the relative importance of using, and the calculation is shown in Appendix 3. In step 5, we need to estimate the improved ranking accuracy by using Eq. 5. As shown in Table 8, the higher score represents a priority and best option. Therefore, steel PPVC achieved the heist score, and it is declared as the best building construction method and materials among alternatives.

Table 6 The results of the individual and generalized surveys

Parameters	60 experts						Total	
	Construction manager		Designer		Academic/Professor			
	Weighting	Priority	Weighting	Priority	Weighting	Priority	Weighting	Priority
Embodied Energy	0.21	3	0.42	1	0.45	1	0.36	1
GHG	0.38	2	0.42	1	0.27	3	0.35	2
LCC	0.41	1	0.16	2	0.29	2	0.29	3
	$CR_a = 0.052 < 0.1$		$CR_a = 0.058 < 0.1$		$CR_a = 0.034 < 0.1$			

Table 7 Initial decision-making matrix

Alternatives	Embodied Energy Table 7 (EE)	GHG (CO_2)	LCC (MYR)
S-PPVC	383.4	21,600	68,040
S-C	395.1	24,300	60,200
C-PPVC	491.4	32,400	61,800
C-C	505.8	35,100	59,000

Table 8 The results of ranking for the multi-attributes analysis WASPAS

Alternatives	Eq. 3. Q1	Eq. 4. Q2	Eq. 5. score	Rank
S-PPVC	0.9615	0.9595	0.9605	1
S-C	0.9447	0.9438	0.9442	2
C-PPVC	0.7911	0.7829	0.7870	3
C-C	0.7783	0.7636	0.7709	4

4.5 Discussion

Choosing the appropriate construction methods and materials when several candidates are available for consideration can be an expensive and time-consuming process for the construction sector. In this research, MADM has been applied in order to compare non-same value parameters. The incentive covering three sustainability criteria is that, some material or system might have a low carbon emission while having a higher cost, which has caused the construction sector to abandon it widely. Finally, in this section, the result of Previous research comparing with the current study. For example, [31] compared the LCA study of conventional versus modular construction methods from cradle-to-gate. The result is consistent with the current research and showed that modular building has less impact in environmental categories than the conventional method. Mao et al. [42] result is consistent with current research and believed that prefabricated has a lower CO_2 emission than conventional building, but highlighted that this could only be achieved by minimizing transportation impacts. Xu et al. [58] revealed that the modular has a higher function in the environmental aspect than traditional construction.

As already discussed in Table 6, the weight result was collected based on expert opinion. Therefore, to overcome human subjectivity in determining the weighting amount for different criteria such as environmental impact, cost, etc., a sensitivity analysis was applied. Weighting has been changed, and criteria were redeveloped for each concrete structure. Hence, all weighing has considered as equal to 0.33 (one divided to three). The result is shown in Table 9, and it's concluded that the human subjectivity has no impact on the final decision by steel PPVC, which is still considered the best option.

Table 9 The results of ranking for the Multi-attributes analysis WASPAS

Alternatives	Equation 3. Q1	Equation 4. Q2	Equation 5. Score	Rank
S-PPVC	0.9557	0.9536	0.9547	1
S-C	0.9464	0.9455	0.9460	2
C-PPVC	0.8005	0.7919	0.7962	3
C-C	0.7911	0.7755	0.7833	4

5 Conclusion

This study accomplished a life cycle sustainability assessment of four different alternatives in the construction industry. In this research, the sustainability assessment of modular and conventional methods for steel and concrete has been described by three criteria: embodied energy, greenhouse gas, and economic aspect. The first criteria are embodied energy that PPVC contributes up to 10% related to the construction in the factory and conventional construction is about 9–12% activity on-site, respectively, for concrete and steel. In conclusion, steel PPVC has a lower embodied energy than alternatives and is considered the best option. The second criteria are GHG that again steel PPVC has a lower emission equal to 20,160 kg CO_2 in the construction stage. The emission related to the construction of PPVC in the factory contributed to 11% and 13%, respectively, for steel and concrete.

In comparison, the emission related to the construction of the site for the conventional method is 28% and 21%, respectively, for steel and concrete. The last stage assessess the cost that the conventional concrete method represents as the most economical method. Finally, the MADM method declared that among all case studies regarding three criteria, steel PPVC is the most sustainable option. Future research might consider some options in their study. This study was based only on the context of Johor, Malaysia. In this manner, results may shift in other nations due to transportation distance. In any case, the discoveries from this research are still essential and contribute to the body of knowledge, as they are the primary considerations to calculate embodied energy, GHG, and cost of sustainability among verity construction materials and methods technique.

Appendix 1

Alternatives	Embodied energy	GHG	LCC
S-PPVC	1.0000	1.0000	0.8671
S-C	0.9704	0.8889	0.9801
C-PPVC	0.7802	0.6667	0.9547
C-C	0.7580	0.6154	1.0000

Appendix 2

Alternatives	Embodied energy	GHG	LCC
S-PPVC	0.3600	0.3500	0.2515
S-C	0.3493	0.3111	0.2842
C-PPVC	0.2809	0.2333	0.2769
C-C	0.2729	0.2154	0.2900

Appendix 3

Alternatives	Embodied energy	GHG	LCC
S-PPVC	1.0000	1.0000	0.9595
S-C	0.9892	0.9596	0.9942
C-PPVC	0.9145	0.8677	0.9866
C-C	0.9051	0.8437	1.0000

References

1. Achal V, Mukherjee A, Kumari D, Zhang Q (2015) Earth-science Reviews Biomineralization for sustainable construction—a review of processes and applications. Earth-Sci Rev 148:1–17. https://doi.org/10.1016/j.earscirev.2015.05.008
2. Ajayi SO, Oyedele LO (2018) Waste-efficient materials procurement for construction projects: a structural equation modelling of critical success factors. Waste Manage 75:60–69. https://doi.org/10.1016/j.wasman.2018.01.025
3. Akasah ZA, Rum NAM (2011) Implementing life cycle costing in Malaysia construction industry: a review. In: Proceeding of International Building and Infrastructure Conference, 7–8 June 2011
4. Alfsen K (2014) Policy Note 1998: 3 The Intergovernmental Panel on Climate Change (IPCC) and scientific consensus How scientists come to say what they say about climate change
5. Alshamrani OS (2015) Life cycle assessment of low-rise office building with different structure–envelope configurations. Can J Civ Eng 43(3):193–200. https://doi.org/10.1139/cjce-2015-0431
6. Ashby M, Johnson K (2013) Materials and design: the art and science of materials selection in product design, Butterworth Heinemann Oxford, vol. 3. p 416
7. Azman MNA, Ahamad MSS, Majid TA, Hanafi MH (2010) The common approach in off-site construction industry. Aust J Basic Appl Sci 4(9):4478–4482
8. BCA S (2009) Design for Manufacturing and Assembly (DFMA) PPVC. Crystal Engineering Corp, pp 1–26
9. Balasbaneh AT, Bin Marsono AK (2017) Proposing of new building scheme and composite towards global warming mitigation for Malaysia. Int J Sustain Eng 10(3). https://doi.org/10.1080/19397038.2017.1293184
10. Balasbaneh AT, Bin Marsono AK, Kasra Kermanshahi E (2018) Balancing of life cycle carbon and cost appraisal on alternative wall and roof design verification for residential building. Constr Innov 18(3):274–300. https://doi.org/10.1108/CI-03-2017-0024
11. Balasbaneh AT, Ramli MZ (2020) A comparative life cycle assessment (LCA) of concrete and steel-prefabricated prefinished volumetric construction structures in Malaysia. Environ Sci Pollut Res. https://doi.org/10.1007/s11356-020-10141-3
12. Balasbaneh AT (2020b) Applying multi-criteria decision-making on alternatives for earth-retaining walls: LCA, LCC, and SLCA. Int J Life Cycle Assess 25:2140–2153. https://doi.org/10.1007/s11367-020-01825-6
13. Balasbaneh AT, Yeoh D, Zainal Abidin AR (2020c) Life cycle sustainability assessment of window renovations in schools against noise pollution in tropical climates. J Build Eng 32(September), 101784. https://doi.org/10.1016/j.jobe.2020.101784
14. Birtles AB (1993) Getting energy efficiency applied in buildings. Energy Environ 4(3):221–252. https://doi.org/10.1177/0958305x9300400302
15. Caruso MC, Menna C, Asprone D, Prota A (2018) LCA-based comparison of the environmental impact of different structural systems. IOP Conf Ser: Mater Sci Eng 442(1). https://doi.org/10.1088/1757-899X/442/1/012010
16. Dong YH, Ng ST (2015) A life cycle assessment model for evaluating the environmental impacts of building construction in Hong Kong. Build Environ 89:183–191. https://doi.org/10.1016/j.buildenv.2015.02.020
17. Energy Commission (Malaysia) (2019) Malaysia Energy Statistics Handbook 2018. Suruhanjaya Tenaga (Energy Comm.), pp 1–86
18. European Committee for Standardization (2011) UNE-EN 15978:2011 Sustainability of construction works - Assessment of environmental performance of buildings - Calculation method. International Standard
19. Fathi MS, Abedi M, Mirasa AK (2012) Construction Industry Experience of Industralised Building System in Malaysia

20. Frischknecht R, Rebitzer G (2005) The ecoinvent database system: a comprehensive web-based LCA database. J Clean Prod 13(13–14):1337–1343. https://doi.org/10.1016/j.jclepro.2005.05.002
21. Goulding JS, Pour Rahimian F, Arif M, Sharp MD (2015) New offsite production and business models in construction: priorities for the future research agenda. Arch Eng Des Manag 11(3):163–184. https://doi.org/10.1080/17452007.2014.891501
22. Guggemos AA, Horvath A (2005) Comparison of environmental effects of steel- and concrete-framed buildings. J Infrastruct Syst 11(2):93–101. https://doi.org/10.1061/(ASCE)1076-0342(2005)11:2(93)
23. Gustavsson L, Joelsson A (2010) Life cycle primary energy analysis of residential buildings. Energy Build 42(2):210–220. https://doi.org/10.1016/j.enbuild.2009.08.017
24. Han MY, Chen GQ, Shao L, Li JS, Alsaedi A, Ahmad B, Guo S, Jiang MM, Ji X (2013) Embodied energy consumption of building construction engineering: case study in E-town, Beijing. Energy Build 64:62–72. https://doi.org/10.1016/j.enbuild.2013.04.006
25. Henkel H-JK (2005) Editorial the revision of ISO standards 14040-3. Int J Life Cycle Assess 10(3):1
26. Heravi G, Nafisi T, Mousavi R (2016) Evaluation of energy consumption during production and construction of concrete and steel frames of residential buildings. Energy Build 130:244–252. https://doi.org/10.1016/j.enbuild.2016.08.067
27. Horváth SE, Szalay Z (2012) Decision-making case study for retrofit of high-rise concrete buildings based on life cycle assessment scenarios. pp 116–124
28. Hwang BG, Shan M, Looi KY (2018) Key constraints and mitigation strategies for prefabricated prefinished volumetric construction. J Clean Prod 183:183–193. https://doi.org/10.1016/j.jclepro.2018.02.136
29. Johnsson H, Meiling JH (2009) Defects in offsite construction: timber module prefabrication. Constr Manag Econ 27(7):667–681. https://doi.org/10.1080/01446190903002797
30. Kamali M, Hewage K (2016) Life cycle performance of modular buildings: a critical review. Renew Sustain Energy Rev 62:1171–1183. https://doi.org/10.1016/j.rser.2016.05.031
31. Kamali M, Hewage K, Sadiq R (2019) Energy & buildings conventional versus modular construction methods : a comparative cradle-to-gate LCA for residential buildings. 204. https://doi.org/10.1016/j.enbuild.2019.109479
32. Kim D (2008) Preliminary life cycle analysis of modular and conventioinal housing in benton harbor, michigan by : Doyoon Kim A practicum submitted in partial fulfillment of requirements for the degree of Master of Science
33. Kohler N, Lützkendorf T (2002) Integrated life-cycle analysis. Build Res Inf 30(5):338–348. https://doi.org/10.1080/09613210110117584
34. Li AS, Ling FYY, Low SP, Ofori G (2016) Strategies for foreign construction-related consultancy firms to improve performance in China. J Manag Eng 32(1):1–6. https://doi.org/10.1061/(ASCE)ME.1943-5479.0000379
35. Li XJ, Zheng YD (2019) Using LCA to research carbon footprint for precast concrete piles during the building construction stage: a China study. J Clean Prod 245:118754. https://doi.org/10.1016/j.jclepro.2019.118754
36. Lim YS, Xia B, Skitmore M, Gray J, Bridge A, Sin Y (2016) Education for sustainability in construction management curricula. 3599 (March). https://doi.org/10.1080/15623599.2015.1066569
37. Lombardi M, Laiola E, Tricase C, Rana R (2017) Assessing the urban carbon footprint: an overview. Environ Impact Assess Rev 66(June):43–52. https://doi.org/10.1016/j.eiar.2017.06.005
38. Lu HR, El Hanandeh A, Gilbert BP (2017) A comparative life cycle study of alternative materials for Australian multi-storey apartment building frame constructions: environmental and economic perspective. J Cleaner Prod 166:458–473. https://doi.org/10.1016/j.jclepro.2017.08.065
39. MacCrimon K (1968) Decision making among multiple attribute alternatives: a survey and consolidated approach. Rand Memorandum, RM-4823-ARPA

40. Malça J, Freire F (2006) Renewability and life-cycle energy efficiency of bioethanol and bio-ethyl tertiary butyl ether (bioETBE): assessing the implications of allocation. Energy 31(15):3362–3380. https://doi.org/10.1016/j.energy.2006.03.013
41. Mao C, Shen Q, Shen L, Tang L (2013) Comparative study of greenhouse gas emissions between off-site prefabrication and conventional construction methods: two case studies of residential projects. Energy Build 66:165–176. https://doi.org/10.1016/j.enbuild.2013.07.033
42. Mao C, Shen Q, Shen L, Tang L (2013b) Comparative study of greenhouse gas emissions between off-site prefabrication and conventional construction methods: two case studies of residential projects. Energy Build 66:165–176. https://doi.org/10.1016/j.enbuild.2013.07.033
43. Monahan J, Powell JC (2011) An embodied carbon and energy analysis of modern methods of construction in housing: a case study using a lifecycle assessment framework. Energy Build 43(1):179–188. https://doi.org/10.1016/j.enbuild.2010.09.005
44. Mourão J, Pedro JB (2007) Sustainable housing: from consensual guidelines to broader challenges. Portugal SB 2007—Sustainable Construction, Materials and Practices: Challenge of the Industry for the New Millennium, May 2007, pp 27–34
45. Pan W, Sidwell R (2011) Demystifying the cost barriers to offsite construction in the UK. Constr Manag Econ 29(11):1081–1099. https://doi.org/10.1080/01446193.2011.637938
46. Peyroteo A, Silva M, Jalali S (2007) Life cycle assessment of steel and reinforced concrete structures: a new analysis tool. Portugal SB 2007—sustainable construction, materials and practices: challenge of the industry for the new millennium, pp 397–402
47. Piroozfar P, Pomponi F, El-Alem F (2019) Life cycle environmental impact assessment of contemporary and traditional housing in Palestine. Energy and Buildings 202:109333. https://doi.org/10.1016/j.enbuild.2019.109333
48. Quale J, Eckelman MJ, Williams KW, Sloditskie G, Zimmerman JB (2012) Construction matters: comparing environmental impacts of building modular and conventional homes in the United States. J Ind Ecol 16(2):243–253. https://doi.org/10.1111/j.1530-9290.2011.00424.x
49. Rosenow J, Eyre N, Bürger V, Rohde C (2013) Overcoming the upfront investment barrier—comparing the German CO_2 building rehabilitation programme and the British green deal. Energy Environ 24(1–2):83–103. https://doi.org/10.1260/0958-305X.24.1-2.83
50. Saaty TL (2008) Decision making with the analytic hierarchy process. Int J Serv Sci 1:83–98
51. Shinde R, Darade MM (2018) Comparison of prefabricated modular homes and traditional R.C.C homes. Int Res J Eng Technol 05(05) | May-2018
52. Tavares V, Lacerda N, Freire F (2019) Embodied energy and greenhouse gas emissions analysis of a prefabricated modular house: The "Moby" case study. J Clean Prod 212:1044–1053. https://doi.org/10.1016/j.jclepro.2018.12.028
53. Teng Y, Li K, Pan W, Ng T (2018) Reducing building life cycle carbon emissions through prefabrication: evidence from and gaps in empirical studies. Build Environ 132(October 2017), 125–136. https://doi.org/10.1016/j.buildenv.2018.01.026
54. The Malaysian electricity generation (2014a) Fossil fuel energy consumption
55. Thormark C (2006) The effect of material choice on the total energy need and recycling potential of a building. Build Environ 41(8):1019–1026. https://doi.org/10.1016/j.buildenv.2005.04.026
56. Treloar GJ (2007) Environmental assessment using both financial and physical quantities. In: 41st annual conference of the architectural science association, Geelong, Australia, November 2007; ANZAScA: Geelong, Australia, 2007, pp 247–255
57. Xing S, Xu Z, Jun G (2008) Inventory analysis of LCA on steel- and concrete-construction office buildings. Energy Build 40(7):1188–1193. https://doi.org/10.1016/j.enbuild.2007.10.016
58. Xu Z, Zayed T, Niu Y (2019) Comparative analysis of modular construction practices in mainland China, Hong Kong and Singapore. J Clean Prod 245https://doi.org/10.1016/j.jclepro.2019.118861
59. Zavadskas EK, Turskis Z, Antucheviciene J, Zakarevicius A (2012) Optimization of weighted aggregated sum product assessment. Elektronika Ir Elektrotechnika 122(6):3–6. https://doi.org/10.5755/j01.eee.122.6.1810

60. Zhai X, Reed R, Mills A (2014) Factors impeding the offsite production of housing construction in China: an investigation of current practice. Constr Manag Econ 32(1–2):40–52. https://doi.org/10.1080/01446193.2013.787491
61. Zhang X, Su X, Huang Z (2007) Comparison of LCA on steel-and concrete-construction office buildings: a case study. In IAQVEC 2007 Proceedings of the 6th international conference on indoor air quality, ventilation and energy conservation in buildings: sustainable built environment, vol 3, pp 293–301